Thomas Armbruster, Rosa Micaela Danisi (Eds.)
Highlights in Mineralogical Crystallography

Also of interest

Modern X-Ray Analysis on Single Crystals
P. Luger, 2014
ISBN 978-3-11-030823-5, e-ISBN (PDF) 978-3-11-030828-0,
e-ISBN (EPUB) 978-3-11-037061-4

Environmental Mineralogy and Bio-Geochemistry of Arsenic
R. Bowell, C. Alpers, H. Jamieson, K. Nordstrom, J. Majzlan (Eds.), 2014
ISBN 978-0-939950-94-2, e-ISBN (PDF) 978-1-61451-797-9,
e-ISBN (EPUB) 978-1-61451-992-8

Spectroscopic Methods in Mineralogy and Material Sciences
G. Henderson, D. Neuville, R. Downs (Eds.), 2014
ISBN 978-0-939950-93-5, e-ISBN (PDF) 978-1-61451-786-3,
e-ISBN (EPUB) 978-1-5015-1061-8

Zeitschrift für Kristallographie – Crystalline Materials
E. Antipov et al. (Editors)
ISSN 2194-4946, e-ISSN 2196-7105

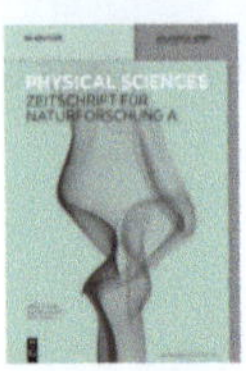

Zeitschrift für Naturforschung A:
A Journal of Physical Sciences
M. Holthaus (Editor-in-Chief)
ISSN 0932-0784, e-ISSN 1865-7109

Zeitschrift für Naturforschung B:
A Journal of Chemical Sciences
G. Müller (Managing Editor)
ISSN 0932-0776, e-ISSN 1865-7117

Thomas Armbruster, Rosa Micaela Danisi (Eds.)

Highlights in Mineralogical Crystallography

DE GRUYTER

Physics and Astronomy Classification Scheme 2010
82.33.Pt, 61.68.+n, 61.05.C-, 61.66.Fn, 91.60.Ed, 81.10.Aj, 31.15.A-, 91.35.-x, 96.30.Za, 71.23.Ft

Editors
Prof. Dr. Thomas Armbruster
Universität Bern
Institut für Geologie
Mineralogische Kristallographie
Freiestr. 3
3012 Bern
Switzerland
armbruster@krist.unibe.ch

Dr. Rosa Micaela Danisi
Universität Bern
Institut für Geologie
Mineralogische Kristallographie
Freiestr. 3
3012 Bern
Switzerland
rosamicaela.danisi@libero.it

ISBN 978-3-11-041704-3
e-ISBN (PDF) 978-3-11-041710-4
e-ISBN (EPUB) 978-3-11-041721-0
Set-ISBN 978-3-11-041711-1

Library of Congress Cataloging-in-Publication Data
A CIP catalog record for this book has been applied for at the Library of Congress.

Bibliographic information published by the Deutsche Nationalbibliothek
The Deutsche Nationalbibliothek lists this publication in the Deutsche Nationalbibliografie; detailed bibliographic data are available on the Internet at http://dnb.dnb.de.

Typesetting: PTP Protago-TEX-Production, Berlin
Printing and binding: CPI books GmbH, Leck
♾ Printed on acid-free paper
Printed in Germany

www.degruyter.com

Editorial

The foundation of modern mineralogical crystallography goes back to the monograph by William Lawrence Bragg "Crystal structure of minerals" published in 1937. This book covers the first collection of crystal structures determined by X-ray diffraction. The 70th anniversary of this new perception in mineralogy was celebrated by a special issue of Zeitschrift für Kristallographie, Vol. 223, issue 1–2 (2008) entitled "Mineralogical Crystallography" comprising review papers by prominent scientists in the field. Since then the subject has further evolved applying new techniques, theoretical approaches, and looking at new kinds of natural inorganic materials.

The more recent 100th anniversary in 2014 of crystallography, proclaimed by the United Nations as International Year of Crystallography, defined as one of the goals to increase the public awareness of the science of crystallography. Admittedly, the structural study of building blocks of life (e.g., proteins) seems recently to dominate crystallographic efforts. However, at the beginning of the development was the study of mineral structures and there are still many open important questions in this field.

As a discipline originally based in geoscience, mineralogical crystallography underwent several modifications varying in a cultural interrelation. In some countries and universities mineralogical crystallography has developed to "Inorganic Material Science" and the direct link to geology faded. In other research centers mineralogical crystallography "still" devotes to the origin, development, and composition of the Earth and our planetary system. In such a focusing, a structural interpretation of minerals as building units of rocks comes to the fore. However, this does not mean that research is only concentrated on geomaterials. Nonetheless, comparison and relation to synthetic materials with similar structures and properties is an important view beyond the horizon. There is an omnipresent relationship and gain in knowledge from minerals to materials and vice versa.

Recent research (Hazen *et al.*, 2008) emphasized the aspect of evolution of minerals and it became clear that there is an obvious link between biological processes and surface mineralogy starting in the Paleoproterozoic (2.2–2 Ga) with the "great oxidation event" leading to multicellular live and skeletal biomineralization. Thus, "biominerals" transformed to Earth's surface mineralogy. Finally, in the Phanerozoic Era (0.57 Ga to present) biology began to dominate the mineralogical diversification of the Earth's surface. Interestingly, the concept of mineral evolution has been developed starting from the question whether there were minerals with "catalytic" properties (e.g., clays and zeolites) already before the origin of life enabling synthesis of simple organic ("biological") molecules. If it is true that to some degree "biology" started on the surface of minerals, the structural relationship of the living and inanimate world may be much more related.

To broaden the field of vision of this book "Highlights in Mineralogical Crystallography" we decided to invite "young" scientists for providing their view of the highlights. Of course we were aware that biological age is not a meaningful parameter if we consider innovative capability. When planning this strategy we had in mind to support the upcoming generation of leading scientists offering them a window for their specific perspective.

The book consists of a series of articles, which cover a wide horizon of topics having no strong connection to each other. As introduced in **Chapter 1,** the scientific community strongly relies on databases. The combination of Raman spectroscopy with diffraction techniques and chemical composition is the key of the large-scale success of the free of charge RRUFF database. In **Chapter 2**, the general trend of mineral evolution has been shown to correspond to an increase of diversity and complexity of mineral structures. The theory behind *ab initio* computer modeling based predictions of mineral surfaces is decrypted in **Chapter 3**. Surface-related properties of minerals are gaining more and more attention in the last decade, although mineralogical crystallography mainly focuses on the characterization of the bulk structure. **Chapters 4 and 5** illustrate two recent groundbreaking discoveries in the field of mineralogy: natural quasicrystals and terrestrial ringwoodite. The identification of icosahedrite in a meteorite shows that quasicrystals are energetically stable states of matter, robust as crystals. In case of terrestrial ringwoodite, the significant presence of water (in terms of OH groups) has substantial impact on mantle geophysics and geochemistry. **Chapter 6** describes the investigation of nanocrystalline bio-related minerals. When conventional X-ray techniques fail to experimentally explore the crystal structure, automated electron diffraction appears to be a powerful method for investigation of submicrometric crystals. At the end, the case of mayenite (**Chapter 7**) illustrates the strong link between mineralogy and material science. The microporous character of this mineral group allows the use of mayenite as catalyst, transparent conducting oxide or crystalline electride.

University of Bern, Switzerland Thomas Armbruster and Rosa Micaela Danisi

Contents

Editorial — V

List of contributing authors — XI

Barbara Lafuente, R. T. Downs, H. Yang, and N. Stone
1 **The power of databases: The RRUFF project — 1**
1.1 Introduction — 1
1.2 A brief history of the RRUFF project — 2
1.3 The RRUFF database — 5
1.3.1 Sample collection and characterization — 5
1.3.2 Reference library — 13
1.3.3 Search and access to mineral records — 14
1.3.4 Software design and server infrastructure — 14
1.4 Experimental observations during collection of Raman spectra — 15
1.4.1 Thermal effects of laser power — 15
1.4.2 Effects of the incident laser wavelength — 16
1.4.3 Fluorescence — 17
1.4.4 Presence of inversion center — 19
1.4.5 Metamict minerals — 20
1.5 Applications of the RRUFF project — 21
1.5.1 Crystal-chemistry analysis of Mars samples — 24
1.5.2 The Mineral Evolution database — 24
1.6 Future directions — 24

Sergey V. Krivovichev
2 **Structural complexity of minerals and mineral parageneses: Information and its evolution in the mineral world — 31**
2.1 Introduction — 31
2.2 Measuring structural complexity — 32
2.2.1 Parsimony and its indices — 32
2.2.2 Information-based complexity measures — 34
2.2.3 Algorithmic complexity — 39
2.3 Information-based structural complexity of minerals: Statistics and basic trends — 48
2.4 Structural complexity and scalar hierarchy — 50
2.5 Structural complexity and the 'ease of crystallization' — 53
2.5.1 General notes — 53
2.5.2 Silica polymorphs — 54
2.5.3 Feldspar compositions — 55

2.5.4 Serpentine-group minerals — 56
2.5.5 Micas and clay minerals — 57
2.5.6 Metastability and structural complexity — 57
2.6 Microevolution of structural complexity in natural processes — 58
2.6.1 Information and chemical reactions: oxidation of pyrite and the cascade of iron sulfate hydrates — 58
2.6.2 Information and temperature: Natural zeolites — 61
2.7 Macroevolution of structural complexity of minerals: Preliminary musings — 65
2.8 Information storage and processing in minerals *versus* living matter — 66
2.9 Conclusions — 67

Sergey V. Churakov

3 Ab initio simulations of mineral surfaces: Recent advances in numerical methods and selected applications — 75

3.1 Introduction — 75
3.2 Theoretical background and simulation methods — 76
3.2.1 Structural rearrangements and surface symmetry — 76
3.2.2 Stability of surfaces of simple ionic solids — 78
3.2.3 Influence of the surface on the electronic structure — 79
3.2.4 Theory of atomistic simulations — 80
3.2.5 Thermodynamics of mineral surfaces — 87
3.2.6 Acid-base properties of surfaces — 88
3.2.7 Simulation packages — 90
3.3 Case studies — 91
3.3.1 Diamond — 91
3.3.2 Rutile — 92
3.3.3 Pyrite — 94
3.3.4 Magnetite — 96
3.3.5 Phyllosilicates — 98
3.4 Outlook — 99

Luca Bindi and Paul J. Steinhardt

4 Natural quasicrystals: A new frontier in mineralogy and its impact on our understanding of matter and the origin of the solar system — 109

4.1 Prelude to quasicrystals — 109
4.2 Revolution in Crystallography: Quasicrystals (1984) — 110
4.3 Revolution in Mineralogy: Natural Quasicrystals (2009) — 112
4.3.1 Extraordinary evidence — 116
4.3.2 Impact — 120
4.4 Outlook — 123

Fabrizio Nestola
5 Ringwoodite: its importance in Earth Sciences — 127
5.1 History of ringwoodite — 127
5.2 Crystal structure and Mg/Fe substitution — 129
5.3 The effect of water on the crystal structure of ringwoodite — 132
5.4 Ringwoodite stability field — 134
5.5 Thermo-elastic properties of ringwoodite — 135
5.5.1 Bulk modulus — 135
5.5.2 The effect of water on the bulk modulus of ringwoodite — 136
5.5.3 Thermal expansion of anhydrous and hydrous ringwoodite — 139
5.5.4 Thermo-elastic properties of ringwoodite: implications for Earth Sciences — 140
5.6 Ascent of diamond-bearing kimberlite magma — 141
5.7 Ringwoodite and the Earth's storage water capacity — 142

Enrico Mugnaioli
6 Investigation of bio-related minerals by electron-diffraction tomography: Vaterite, dental hydroxyapatite, and crystalline nanorods in sponge primmorphs — 149
6.1 Introduction — 149
6.2 Automated diffraction tomography (ADT) — 151
6.3 Experimental — 154
6.3.1 Conventional TEM, ADT and structure analysis — 154
6.3.2 Sponge primmorphs sample preparation — 154
6.3.3 Vaterite synthesis — 154
6.3.4 Human hydroxyapatite extraction — 155
6.4 Characterization of unknown biominerals inside tissues: Crystalline nanorods in sponge primmorphs — 155
6.5 Ab-initio structure determination of complex bio-related nanocrystalline minerals: Vaterite — 157
6.6 Structure investigation of human calcified tissues: Dental hydroxyapatite — 159
6.7 Conclusions and perspectives — 161

Frank Gfeller
7 Mayenite $Ca_{12}Al_{14}O_{32}[X^{2-}]$: From minerals to the first stable electride crystals — 169
7.1 The discovery and chemical composition of mayenite $Ca_{12}Al_6Al_8O_{32}[O]$ — 169
7.2 The crystal structure of mayenite and its relation to the garnet structure — 170
7.3 Extra framework anions and structure distortion — 170

7.3.1 Centre, off-centre or both? — 173
7.3.2 Extraframework occupants in “dry” mayenite — 175
7.3.3 Hydroxylated mayenite — 176
7.4 Zeolite or anti-zeolite? — 178
7.5 Applications of mayenite ceramics — 179
7.5.1 Mayenite the first thermally and chemically stable electride — 179
7.5.2 Active oxygen O^-, O_2^- in mayenite as support of syngas catalysis — 180
7.5.3 Cation and anion doping of mayenite — 181
7.6 The mineral mayenite — 181
7.6.1 The mayenite supergroup — 182
7.6.2 Achtarandite — 185
7.6.3 Silicates with mayenite-type structure — 187
7.7 Synthesis and structure of wadalite-group members — 187
7.8 Outlook — 190

Index — 197

List of contributing authors

Frank Gfeller
Mineralogical Crystallography
Institute of Geological Sciences
University of Bern
Freistr. 3
3012 Bern
Switzerland
frank.gfeller@krist.unibe.ch

Luca Bindi
Dipartimento di Scienze della Terra
Università di Firenze
Via La Pira 4
50121 Firenze
Italy
luca.bindi@unifi.it

Paul J. Steinhardt
Department of Physics & Princeton Center for Theoretical Science
Princeton University
Princeton
NJ 08540
USA
steinh@princeton.edu

Enrico Mugnaioli
Dipartimento di Scienze Fisiche, della Terra e dell'Ambiente
Università degli Studi di Siena
Via Laterino 8
53100 Siena
Italy
enrico.mugnaioli@unisi.it

Sergey V. Churakov
Mineralogy
Institute of Geological Sciences
University of Bern
Baltzerstrasse 1+3
3012 Bern
Switzerland
sergey.churakov@geo.unibe.ch
and
Laboratory for Waste Management
Paul Scherrer Institute
5232 Villigen-PSI
Switzerland
sergey.churakov@psi.ch

Sergey V. Krivovichev
Department of Crystallography
St. Petersburg State University
University Emb. 7/9
199034 St. Petersburg
Russia
s.krivovichev@spbu.ru

Barbara Lafuente
Department of Geosciences
University of Arizona
1040 E. 4th Street
Tucson
AZ 85721-0077
USA
barbaralafuente@email.arizona.edu

Robert T. Downs
Department of Geosciences
University of Arizona
1040 E. 4th Street
Tucson
AZ 85721-0077
USA
rdowns@u.arizona.edu

Hexiong Yang
Department of Geosciences
University of Arizona
1040 E. 4th Street
Tucson
AZ 85721-0077
USA
hyang@email.arizona.edu

Nate Stone
Open Data Repository
21 Shaker Road
Gray ME 04039
USA
nate.stone@opendatarepository.org

Fabrizio Nestola
Dipartimento di Geoscienze
Università di Padova
Via Gradenigo 6
35131 Padova
Italy
fabrizio.nestola@unipd.it

Barbara Lafuente, R. T. Downs, H. Yang, and N. Stone

1 The power of databases: The RRUFF project

1.1 Introduction

In today's world, many people rely on the availability of databases to perform daily activities such as checking meteorological forecast, finding a recipe, or verifying the spelling of a word. These are examples of actions that anyone can do quickly, routinely, and often at little cost, because of access to the Internet and the development of extensive databases.

The science community has always relied on databases, which in the earlier years were only available through collations of journals, books and data records. The analogic nature of these data resources dramatically limits the process of searching records and establishing relations between datasets. Nowadays, digital scientific databases and the World Wide Web have changed the way we can do science. This new generation of databases not only provides access to information, but allows data sharing between multidisciplinary research projects and make it possible to cross-link diverse types of data.

Minerals are very important to both the world economy and research science, they supply the raw materials necessary for the development of modern societies and their study provides knowledge of our planet, the solar system, and the very nature of the universe. As such, their identification and characterization is critical to many fields. However, only a few experts can accurately identify minerals without analytical instruments and their complementary databases. In general, minerals are uniquely characterized by both chemistry and crystal structure, or their derivative properties. For instance, the derivative property, color, can be a function of chemistry, or crystal habit a function of structure.

The characterization and differentiation of minerals in the 19th century was largely based on chemical composition and crystal forms. This led to the appearances of compendia of minerals and their diagnostic properties in the form of books that become the earliest publicly available databases of minerals, such as Dana's System of Mineralogy (1837–1997) [1], Hintze's Handbuch der Mineralogie (1897–1933) [2], or Goldschmidt's Atlas der Krystallformen (1913–1923) [3] containing drawings of all recognized crystals forms of minerals.

The discovery of X-ray diffraction techniques in 1912 brought about a dramatic change in the way minerals could be characterized because the diffraction pattern of a given mineral species remains consistent without regard to the crystal form, habit, color, or other variability in properties. This has led to a large amount of definitive data that is used to help identify and characterize minerals. Subsequent technological innovations have increased the speed and volume of data acquisition, and therefore, the need of databases to organize and make available all this information.

In addition to chemical and diffraction techniques, spectroscopy has proven to be a valuable derivative methodology used to identify and characterize minerals. In particular, Raman spectroscopy has shown itself to be a competitive technique because Raman spectra of minerals depend fundamentally on their chemistries and structures. Moreover, this technique is generally nondestructive, requires little sample preparation, it is quick and holds the promise of low cost. Infrared spectroscopy is another popular technique that is complementary to Raman and is capable of providing distinctive spectra suitable for search/match routines. Both spectroscopy techniques are undergoing major instrumentation changes towards portable and handheld systems (e.g. [4–7]).

Mineral identification using Raman spectroscopy is normally performed by search/match routines that compare the acquired spectrum with reference spectra from a database. The purpose of the RRUFF project is to develop such a Raman database by measuring the chemistry, X-ray diffraction patterns, Raman, and infrared spectra of the known minerals and to make these data readily and freely available to the scientific community, industry, and the general public. RRUFF database currently contains about 7000 mineral samples representing 3500 mineral species. The power of the RRUFF database is that its data is not collected from disparate publications and procedures but rather is collected with the same methodology, on the same set of samples and with the same instrument.

Although they are not currently included in RRUFF, there are other analytical and spectroscopy techniques that aid in the characterization of minerals [8]. For example, Laser-Induced Break-down Spectroscopy (LIBS) provides chemical composition by measuring all the elements present in a mineral sample. Mössbauer spectroscopy is also extensively used in mineralogy to examine the various valence states and the type of coordination polyhedron occupied by iron atoms.

1.2 A brief history of the RRUFF project

Until recently, Raman spectroscopy was only accessible to specialists because it required complex instruments worth hundreds of thousands of dollars. In 2003, the Hamilton Sundstrand Sensor Systems company (CA, USA) and the University of Arizona (USA) were funded by NASA to develop a miniaturized high performance Raman system for Mars surface studies [9, 10]. The success of this project demonstrated that inexpensive small Raman instruments had the capability to make Raman spectroscopy a popular stand-alone field tool for routine and unambiguous identification of minerals and other materials by non-experts and thus, would eventually become available to the broad public.

Mineral identification using Raman spectroscopy needs reference spectra, so anticipating the increasing use of the technique, Pr. Downs at the University of Arizona built a Raman spectral database for minerals by obtaining representative samples of minerals that are also characterized and identified by X-ray diffraction and chemical composition. There are a couple of models that can be used to fund database development. One is to charge an annual license, such as the funding model used by International Centre for Diffraction Data (ICDD) (http://www.icdd.com/). However, if the cost of Raman instruments become so inexpensive that it is less than the cost of a license, then the ICDD model of funding is not reasonable. The RRUFF project database is based on the model of paying upfront for database development and providing the data for free.

Mr. Michael Scott, the founding president of Apple Computers, provided the funding to begin the project and create a proof of concept. He named it the RRUFF Project [11]. With this funding, the project developed procedures for validation and verification and the purchase of required instruments, including X-ray diffractometers and Raman spectrometers.

A requirement from Mr. Scott was that all the data collected by the RRUFF project would feed an open-access Raman spectral database not only for geologists and mineralogists, but for the general worldwide public. Therefore, a website for the RRUFF database (http://rruff.info) was built to provide free access to all the collected data, allowing users to view and download X-ray diffraction patterns, Raman and infrared spectra, electron microprobe data, references, photographs, and acquisition data of well characterized mineral samples. The RRUFF database is constantly reviewed and updated, following strict protocols to ensure the quality and durability of the data. It currently receives ~80000 hits per week. Its success, together with the interest of Michael Scott in mineralogy and Raman spectroscopy, has led him to also fund a similar free-access database, called WURM (http://www.wurm.info/), which contains the results of ab initio Raman spectra calculations [12].

It was also important in the early days of the RRUFF project to obtain a list of the known mineral species. With the help of the International Mineralogical Association (IMA) and its nomenclature commission, an outreach committee of the IMA was formed to develop and maintain an interactive and searchable list of the IMA-approved minerals. This list, referred in this paper as the IMA-List, is found at: http://rruff.info/ima/ (Fig. 1.1) and provides our preferred interface to the data of RRUFF as well as the Handbook of Mineralogy [13] maintained by the Mineralogical Society of America, the American Mineralogist Crystal Structure Database (AMCSD) for structural data [14–16], a reference library of publications and in some cases their pdf's, a Mineral Evolution database for mineral localities and their ages, and links to other databases that search on mineral names. In addition, and in order to promote private use of the IMA-List of minerals, the site includes an option to create additional links to any other website that will accept queries based on mineral names. For example, we have a link to our mineral museum database. The site also offers a downloadable spell check dictionary file containing the mineral names from the IMA-List which can be added to a user's computer dictionary.

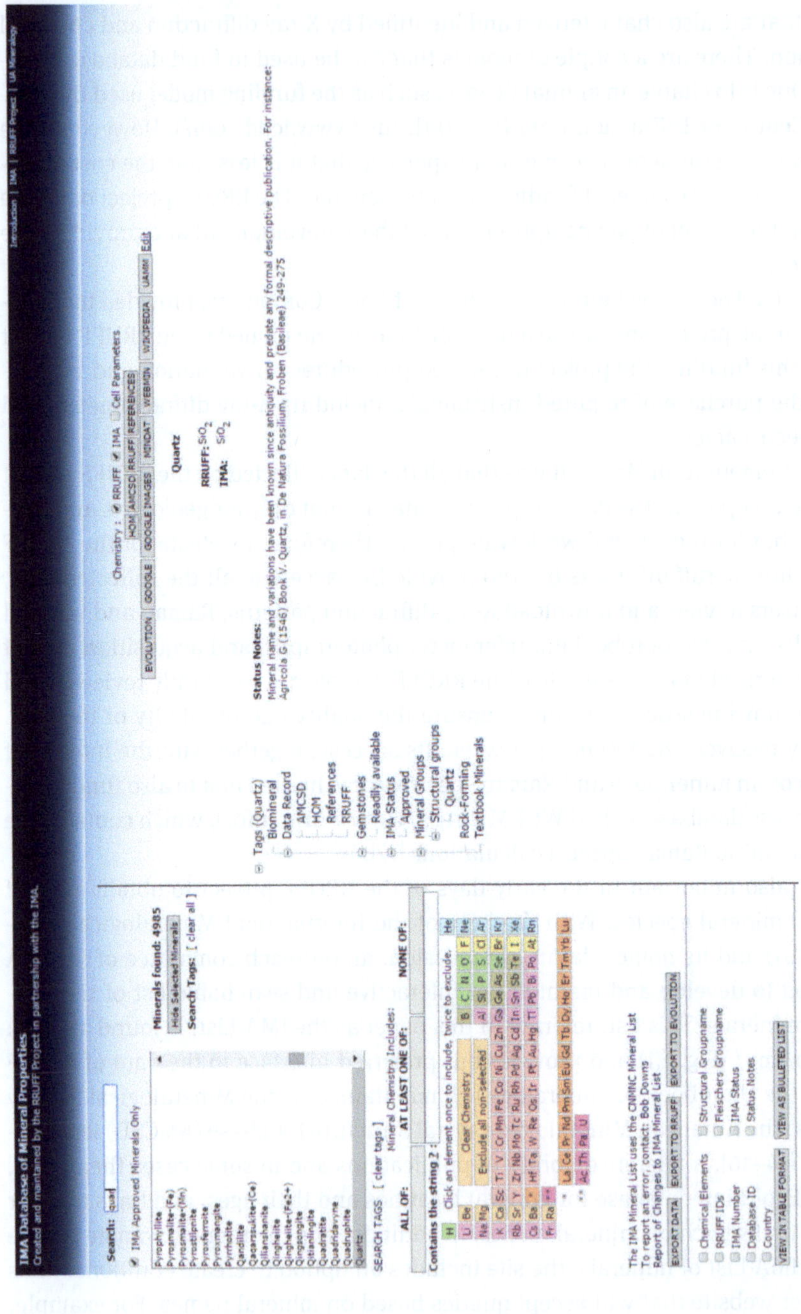

Fig. 1.1: The IMA-List homepage layout (http://rruff.info/ima/). At the top left is the interactive list of IMA-approved minerals, which is searchable by mineral name (e.g. quartz) or by chemistry through the available periodic table. At the top right is the list of provided links to other mineral database, including our customized link to the University of Arizona Mineral Museum (UAMM).

1.3 The RRUFF database

As of this moment, 3527 of the 4985 known mineral species have been incorporated into the RRUFF sample collection. As data from a sample is collected, it is posted into the database with password restricted access, referred to as non-public access. After a review process, if the data appears to be representative of the sample, then the password restriction is removed and the data becomes publicly accessible. As a consequence, measurements from only 2128 mineral species are currently publicly accessible. When possible, data from at least two samples of each species, ideally from different localities, are included in the database. Having multiple samples provides a means to confirm data and capture the chemical variability frequently observed in minerals. For instance, the database currently contains 42 records on the important olivine forsterite-fayalite series, with associated Mg-Fe chemical variations. In total, data from 3791 samples are publicly accessible through the RRUFF database.

1.3.1 Sample collection and characterization

Mineral samples for the database were initially obtained from the University of Arizona Mineral Museum (UAMM) and from the purchase of the Cureton reference collection of 6500 specimens by Michael Scott. Currently, new mineral specimens are mostly obtained from donations by collectors, dealers, and museums around the world. In 2013 alone, the RRUFF project received over 10000 samples through donations, including two major ones: Mr. Rock Currier (~8000) and Dr. Robert Lavinsky (~2000).

We insert the samples of interest into the database by giving them a unique RRUFF ID (e.g. R070563) and we scan and post the original labels, if they exist. We also record important acquisition data, such as source and locality, according to the label. After the record has been created, the next step is to confirm the identity of the sample. We have found that more than 25 % of all the samples that we have studied have been previously incorrectly identified on their labels. This failure rate illustrates the difficulty in determining the identity of a mineral, even for experts. Until the identification is confirmed, users should carefully suspect the correctness of the mineral name. The *Status* line in the top portion of each sample record reports the degree of the certainty in the identification and the techniques used. Even though our goal is to have all samples characterized by means of X-ray diffraction and chemical analysis, due to limited resources and the cost of chemical analysis, we try to identify them first by X-ray diffraction because this equipment is in our laboratory. However, some minerals can be difficult to be unambiguously identified by X-ray diffraction. For example, different mineral species belonging to the garnet or amphibole groups may have similar cell parameters and because they are isostructural, their diffraction patterns will be similar. In these cases, chemical composition is required. When using RRUFF data, it is always important to review the *Status* field in order to know the degree of certainty with which the minerals has been identified.

Most specimens in the RRUFF project are also documented with a *macro* photograph of the entire sample, intended to show host material, and a *micro* photograph to illustrate the quality of the mineral specimen under examination. Examples of both *macro* and *micro* photos are shown in Fig. 1.2.

Fig. 1.2: Photos *macro* (top) and photos *micro* (bottom) of some RRUFF samples. From left to right: sigloite (R050576), bobdownsite (R050109), and xanthoconite (R080013).

1.3.1.1 Chemical analysis

The chemical composition is mainly determined with electron-microprobe analysis (EPMA) complemented, when necessary, with inductively coupled plasma mass-spectrometry (ICP-MS) analysis for the measurement of light elements (lighter than oxygen, such as B, Be, and Li).

Electron-microprobe analyses are conducted at the University of Arizona Electron Microprobe Laboratory at the Lunar and Planetary Laboratory (LPL) on a fully automated CAMECA SX50 or SX100 Ultra instrument. A detailed description of microprobe analysis applied to geology and mineralogy can be found in Reed [17] and the calculations to determine the empirical chemical formula are explained in detail in Deer et al. [18]. The determination of light-elements (H, Li, Be, B) is done by inductively coupled plasma mass-spectrometry (ICP-MS) on a Thermo Scientific X-SERIES 2 Quadrupole Inductively Coupled Plasma Mass Spectrometer (Q-ICPMS).

The chemical analysis for each mineral is posted to the RRUFF website along with a photo of the image of the polished fragment used in the probe. An Excel file con-

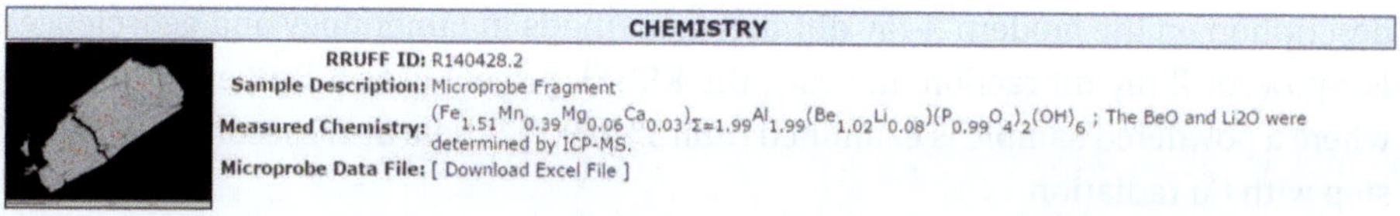

Fig. 1.3: Chemistry section of lefontite (R140428) showing the measured chemistry, photo of SEM image (link to a larger high-resolution image), and the access to the Excel spreadsheet.

taining all the probe data is also provided and it can be downloaded and manipulated by anyone who is interested in the results or to use as a template for other analyses (Fig. 1.3).

The IMA-List provides an interactive interface to search for minerals based on their ideal chemistry using a periodic table (Fig. 1.4). With this tool, the search can be done not only by elemental composition but also by valence state (i.e. Fe^{3+}) and strings (i.e. H_2O, OH), which results in a single selected mineral, or a set of minerals.

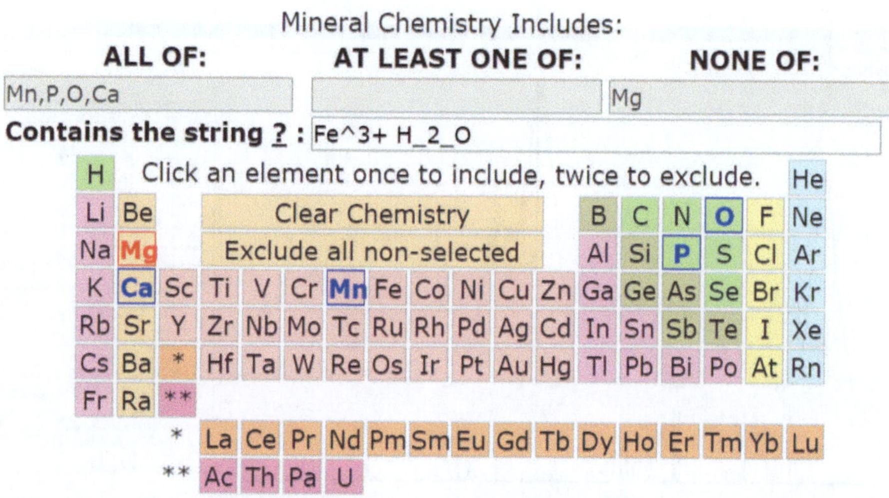

Fig. 1.4: Periodic table interface in the IMA-List. The example illustrates a searching based on minerals that contain the elements Mn, P, O and Ca, the strings Fe^{3+} and H_2O, and exclude Mg. The system retrieves seven minerals meeting these conditions: bederite, jahnsite-(CaMnFe), jahnsite-(CaMnMn), keckite, wicksite, wilhelmvierlingite, and zodacite.

1.3.1.2 X-ray diffraction

All the RRUFF samples are examined by either powder or single-crystal X-ray diffraction. When there is sufficient sample then powder X-ray diffraction is the method of choice because it produces a pattern that matches the established gold standard format for identifying minerals. The reader can refer to Lavina et al. [19] for an extended description on the modern X-ray diffraction methods in mineralogy and geosciences. For powder X-ray diffraction analysis, the RRUFF project uses a Bruker D8 Advance where a powdered sample is examined from 5° to 90° 2-theta at 2.0 seconds per 0.010° step with Cu radiation.

The diffraction patterns are compared against the ICDD using the EVA search/match module provided by Bruker. Each measured diffraction pattern is then indexed and cell parameters are refined with CrystalSleuth [20] using *h k l* values from AMCSD, as computed by the module XPOW [21] included in the XtalDraw software [22]. The cell parameter refinement in CrystalSleuth is based upon determining the positions of isolated diffraction peak $K\alpha 1$ and $K\alpha 2$ doublets and fitting the set of positions with a non-linear least-squares algorithm [20, 22] (Fig. 1.5). The peaks are fit using a χ^2 minimization routine known as the Levenberg-Marquardt Method similar to that described in Press et al. [23].

On the RRUFF website, the diffraction pattern is displayed and the profiles are posted as XY-ASCII files, along with the refined crystallographic parameters.

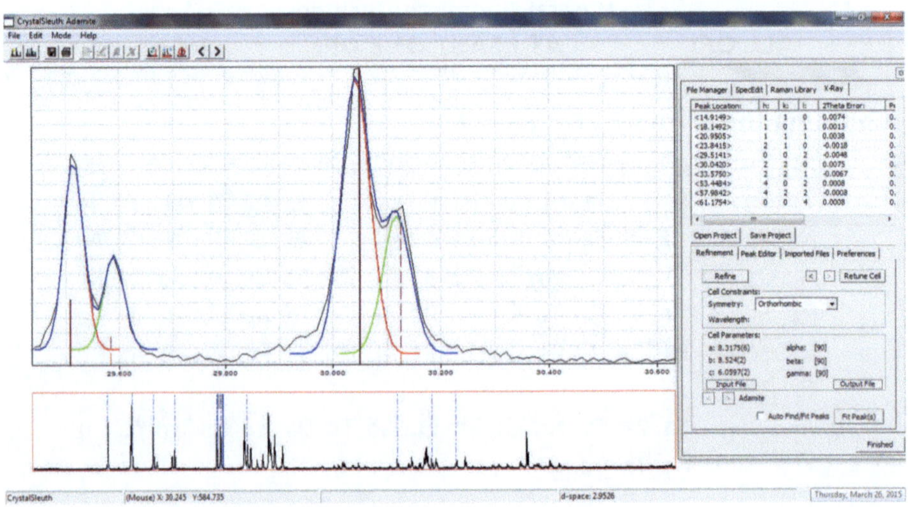

Fig. 1.5: CrystalSleuth interface showing an example of fit and index processes of an experimental powder X-ray diffraction pattern of an adamite sample.

If the identification or refinement by powder X-ray diffraction is ambiguous due to, for instance, extreme peak overlap, or the amount of sample is too small for powder work, then single-crystal X-ray diffraction is the preferred technique. The analyses are conducted using a state-of-the-art Bruker X8 diffractometer with Apex2 CCD detector and Mo radiation.

In this case, the mineral is identified on the basis of its cell parameters and search/match is performed using the IMA-List, which offers a routine for search-

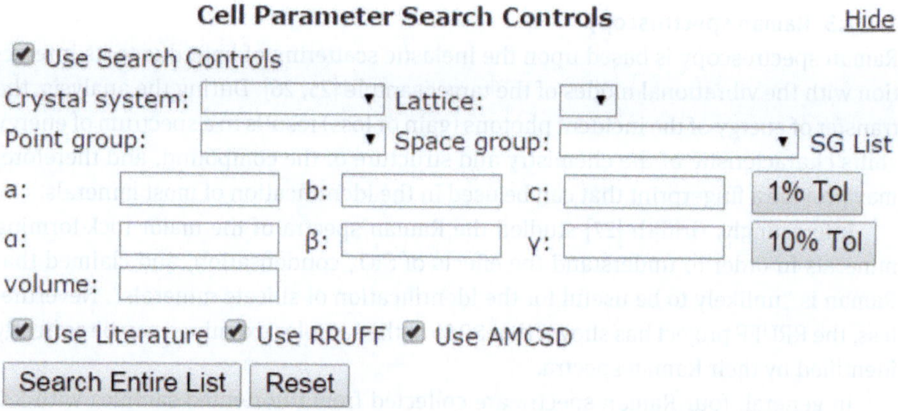

Fig. 1.6: Cell parameter search control interface in the IMA-List.

ing cell parameter extracted from the literature and from the RRUFF and AMCSD databases. It also provides a tolerance of 1 %, 10 %, or any value chosen by the user to the cell parameters to enable a search within a range of values (Fig. 1.6).

With single-crystal diffraction data, a theoretical powder diffraction pattern is computed using XPOW (Fig. 1.7) and it is posted in the database along with its associated XY-ASCII file.

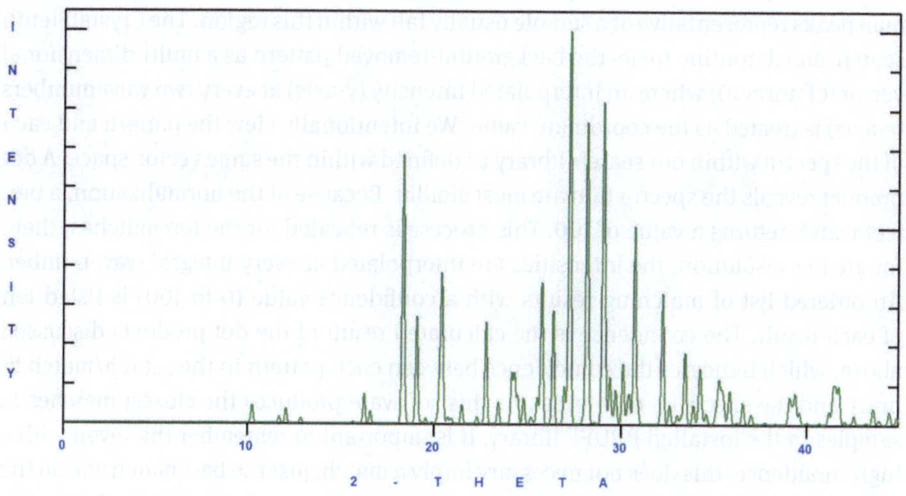

Fig. 1.7: Theoretical powder X-ray diffraction pattern of iranite [24] generated by XPOW.

1.3.1.3 Raman spectroscopy

Raman spectroscopy is based upon the inelastic scattering of light due to its interaction with the vibrational modes of the target sample [25, 26]. During the analysis, the transfer of energy of the incident photons (gain or loss) results in a spectrum of energy shifts characteristic of the chemistry and structure of the compound, and therefore, may provide a fingerprint that can be used in the identification of most minerals.

Interestingly, Griffith [27] studied the Raman spectra of the major rock-forming minerals in order to understand the effects of SiO_4 condensation, and claimed that Raman is "unlikely to be useful for the identification of silicate minerals". Nevertheless, the RRUFF project has shown that 80 % of the samples examined can be correctly identified by their Raman spectra.

In general, four Raman spectra are collected from unoriented samples with 532 (green) and 785/780 (red) nm lasers in a Thermo-Almega micro Raman system. The instrument is equipped with an Olympus BX microscope with 10X, 50X, and 100X objective lenses. Whenever possible, the measurements are collected using 50X. The lasers are partially polarized with 4 cm^{-1} of maximum resolution and a spot size < 1 μm. High-resolution narrow scans, usually from 70–1400 cm^{-1}, and low-resolution wide scans usually from 70–4000 or 70–6500 cm^{-1} are collected with the 532 and 785/780 nm lasers from exactly the same spot in the sample and until the signal to noise ratio is optimal.

The high-resolution narrow scans are used as reference spectra in the search/match library installed with CrystalSleuth (Fig. 1.8), because the strongest main Raman peaks representative of a sample usually fall within this region. The CrystalSleuth search/match routine treats the background-removed pattern as a multi-dimensional vector (of norm 10) where an interpolated intensity (y-axis) at every two wavenumbers (x-axis) is treated as the coordinate value. We intentionally view the pattern and each of the spectra within our search-library as defined within the same vector-space. A dot product reveals the spectra that are most similar. Because of the normalization, a perfect match returns a value of 100. This process is repeated for the top matches; then, for greater resolution, the intensities are interpolated at every integral wavenumber. An ordered list of matching results with a confidence value (0 to 100) is listed left of each result. The confidence is the calculated result of the dot products discussed above, which indicates the coincidence between each pattern in the search/match library and the spectrum of interest. As this software produces the closest matches to samples in the installed RRUFF library, it is important to remember that even with a high confidence, this does not necessary imply a match, just the best match within the database. In some cases, it is possible that the right reference is not even in the library.

The low-resolution wide scans are performed with both 532 and 785 nm lasers in order to record other Raman peaks, such as the O-H stretching modes, and other spectral artifacts. With our instrument, only the spectra collected with the 532 nm laser reveal the presence of O-H bands that are often located in the 2700–3700 cm^{-1} region, as

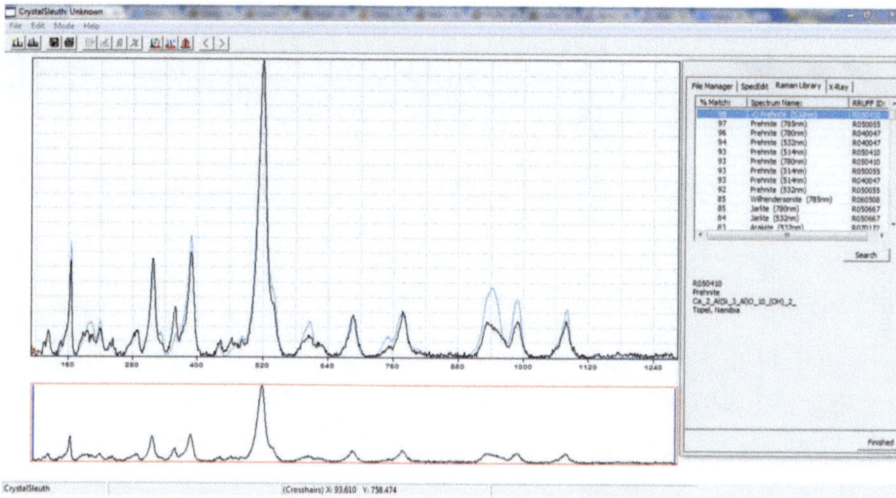

Fig. 1.8: CrystalSleuth interface showing an experimental unknown Raman spectra (in black), which matches the spectrum of the RRUFF record for prehnite R050410 (in blue).

shown in Fig. 1.9. The sensitivity of the silicon detector in the 785 nm laser is quite poor at higher wavenumbers which makes it useless for the detection of the O-H modes.

CrystalSleuth is used to correct the spectral baseline, using a subroutine from the Razor library (http://www.spectrumsquare.com) by Spectrum Square Associates, to remove cosmic rays events from patterns [28], to trim edges, to reverse X-axis display, and to visualize and compare multiple spectra. Both raw and processed data are included on RRUFF as XY-ASCII files.

The intensities of certain Raman peaks of a mineral may vary appreciably as a function of sample orientation, depending upon its symmetry and the origin of the Raman modes (Fig. 1.10).

Such variations in intensity can significantly affect the success of the search/match routines. Thus, in certain cases, especially for rock-forming minerals, Raman spectra collected in additional different orientations with respect to the polarization direction of the incident laser are collected. The analyses are conducted with an IRIS Raman instrument built by Alex Goncharov and Victor Struzhkin of the Geophysical Laboratory at the Carnegie Institution of Washington. It uses a tunable 100 mW Ar-ion laser (usually at 514.532 nm) and a Jobin Yvon Spex HR 460 spectrometer equipped with a liquid nitrogen cooled Princeton Instruments 1152 × 256 pixel CCD detector. We use a 1200 groves mm^{-1} grating centered at 530.4 nm and collect data with Roper Instruments Winspec/32 software. Samples can be oriented onto a goniometer stage and rotated precisely into position, or spun at speeds up to 720 degrees per sec.

Some mineral records also provide a pdf file with the description of the Raman active modes corresponding to the mineral species (e.g. beryl, R040002). The analysis

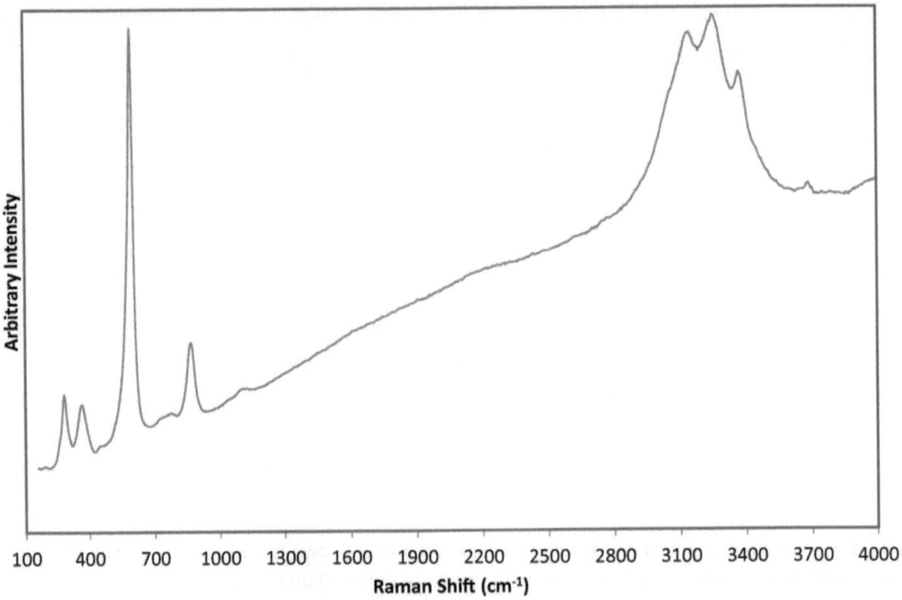

Fig. 1.9: Low-resolution Raman spectrum of tetrawickmanite, $Mn^{2+}Sn^{4+}(OH)_6$ (R100003), collected with the 532 nm laser, illustrating the O-H stretching vibrations Raman bands between 2700 and 3700 cm^{-1}.

is done by using the crystal structure data from AMCSD and the SAM routine at the Bilbao Crystallographic server (http://www.cryst.ehu.es) [29].

1.3.1.4 Infrared spectroscopy

Similarly to Raman spectroscopy, the infrared spectrum can provide additional information and a quick way to identify a mineral, assuming that a database is available for comparison. The two techniques are complementary because while Raman scattering requires a change in polarizability with vibration, infrared requires a change in dipole moment. Therefore, in molecules with different elements of symmetry, certain bands may be infrared active, Raman active, both or neither. When sufficient powder sample exists, the infrared absorption spectra are collected at the California Institute of Technology (Caltech) in Dr. George Rossman's lab. The analyses are performed using a SensIR Durascope on a Thermo Nicolet Magna 860 FTIR (wavelength 4000–375 nm). The collected data is included as XY-ASCII file. The RRUFF project does not include a search/match routine for identification of minerals by infrared spectroscopy. Additional details can be found at George Rossman's website at Caltech. The user can access the Mineral Spectrosocopy Server (http://minerals.gps.caltech.edu/) where visible, near-infrared, and infrared absorption spectra are available for a number of min-

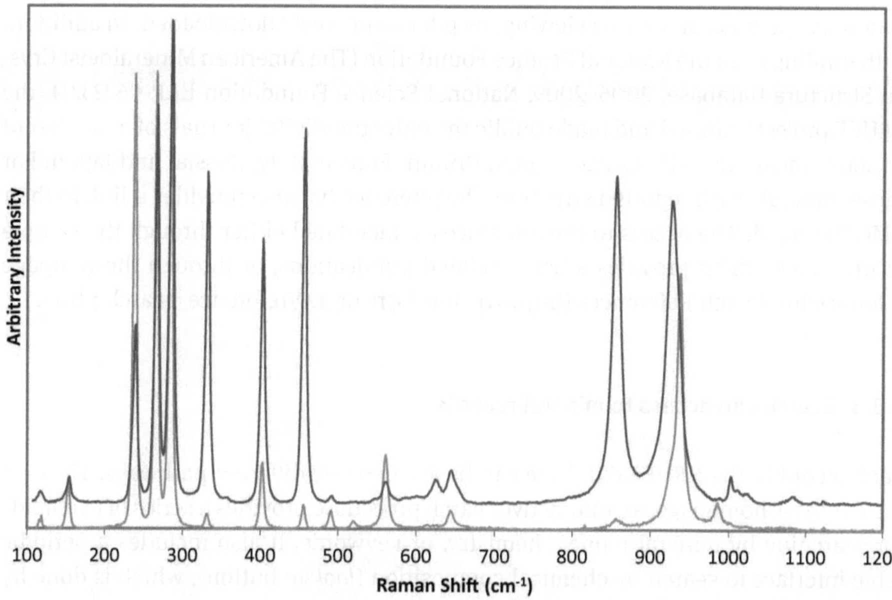

Fig. 1.10: Raman spectra of topaz (R040121) in two different orientations. The lower spectrum (orange) was recorded with the laser parallel to the *b*-axis, polarized parallel to *c*, while in the upper spectrum (blue) the laser is parallel to the *c*-axis, polarized parallel to *b*.

erals, including rock-forming minerals, gem minerals, and other minerals of particular interest.

1.3.1.5 Data from external sources

Some rare minerals are very difficult to obtain by the RRUFF project. However, if the Raman and infrared spectra are available in the literature, we contact the authors and invite them to share their data through the RRUFF database. Microprobe analysis and crystal structure information, if available from the manuscript, are also included in the record. We thank all the authors for sharing their data with the RRUFF project. The list of minerals needed to complete the project can be found at http://rruff.info/about/minerals_needed.php. If you have Raman or infrared spectra of minerals that are not currently represented in the database, we encourage you to contact us.

1.3.2 Reference library

The RRUFF project includes an extensive reference library of publications directly linked to their associated minerals. For the most part, these articles are focused on spectroscopy, structure, and chemistry of minerals. The complete list of collaborators can be found at: http://rruff.info/about/about_publishers.php. Only when the publication has permission for free viewing, its pdf can be freely downloaded. In addition, with funding from the National Science Foundation (The American Mineralogist Crystal Structure Database, 2006–2009, National Science Foundation EAR-0622371), the RRUFF project scanned and made public the older pre-digital journals of a number of societies including USA, Canada, Great Britain, France, Italy, Russia, and Japan. For those minerals with reported structure, the reference list also provides a link to their AMCSD record. The access to the references is facilitated either through the sample record itself, which provides a list of related publications, or through the webpage interface for Search References (http://rruff.info/rruff_1.0/reference_search.php).

1.3.3 Search and access to mineral records

Each record in the RRUFF database can be accessed by different pathways. Through the database homepage, an interactive search procedure provides a series of text fields for searching by mineral name, chemistry, or keywords. It also includes a periodic table interface to search by chemical composition (*lookup* button), which is done by selecting elements included in the chemistry (click element once), or elements that needs to be excluded from the chemistry (click element twice). For example, all silicate minerals can be obtained by selecting only Si and O. To obtain the SiO_2 polymorphs, the rest of elements in the periodic table needs to be excluded. The *Exclude all non-selected* button excludes all of them automatically. The search hits can be sorted and displayed in different formats that can be set as default.

The mineral records can be also accessed by typing the name of the mineral or the RRUFF ID in the website URL in the format: http://rruff.info/[*mineral name or RRUFF ID*]. For example, rruff.info/quartz returns a page listing all the public quartz samples that have been analyzed by the RRUFF project, while rruff.info/R040031 displays all the data collected for one of the specific samples of quartz. This type of access also facilitates a way for other websites or software to independently access data from the RRUFF webpages. The IMA-List uses the URL method to provide a link to the mineral records in RRUFF. This link will be only active (black font) when the RRUFF database contains at least a record of the selected mineral.

1.3.4 Software design and server infrastructure

The RRUFF database is built using the open-source database-management software MySQL, which runs on all platforms. MySQL is combined with the PHP scripting language to create the database structure, to populate and maintain the database, and to support data retrieval. PHP not only securely supports the MySQL database but also it is the link between the database and the Web page.

The Web search interface is dynamically created with PHP using the *Select* commands to access the criterion tables in the MySQL database. We use an Apache Web server running on a Linux platform, which is well suited for PHP and MySQL. The searchable fields in the user interface include interactive HTML elements such as text boxes, drop-down list boxes, radio buttons, push buttons, etc., which are created with PHP. The results of the queries populate the search criteria fields in the HTML form.

The RRUFF project spends considerable time and resources to ensure the availability of the RRUFF database at all times. As the database is used worldwide, it is a priority for the project to provide users with a reliable tool that can be used at any time and whose data last over time. Server infrastructure used for the RRUFF database uses RAID (Redundant Array of Inexpensive Disks) for local storage to ensure data integrity within the local archive. This ensures that a single or double disk failure in the system will not result in loss of data. This hardware is regularly maintained and upgraded to ensure performance and availability.

1.4 Experimental observations during collection of Raman spectra

An outcome of building the RRUFF database is the opportunity to examine hundreds of minerals by Raman spectroscopy and to observe diverse factors that may affect the collection of the data [30]. We summarize some of these factors in this section.

1.4.1 Thermal effects of laser power

During the collection of a Raman spectrum, the incident laser beam may heat the illuminated portion of the sample, which, in some cases, can cause phase transitions, loss of hydration water, or partial or total decomposition of the sample. The increase of the temperature depends upon experimental conditions, such as excitation wavelength, laser power, and exposure time, as well as mineral properties, including size, chemistry, and color. When heating effects are suspected, we typically lower laser power, modify exposure time, or defocus the laser beam. The downside of reducing the laser power to its minimum value is that it also implies lowering the intensity of the Raman signal and worsening the signal to noise ratio, therefore providing a lesser quality Raman spectra.

Usually small or thin isolated crystals tend to burn more easily because they are not able to dissipate much of the heat. When we collect Raman data from the same single crystal used for X-ray diffraction analysis, the power of the laser is usually reduced to its lowest value. Minerals with weak bonds, such as many of the sulfides or highly hydrated minerals, also tend to burn more easily even under the lowest laser intensity setting. Cinnabar (HgS) provides an example where the high intensity 514 nm laser produces the spectrum of sulphur (S). Opaque phases, such as iron oxides, absorb more of the incident radiation making them also more susceptible to burning. Interestingly, even though the 532 nm laser is higher energy than the 785 nm laser, we have found cases where a mineral burns under 785 nm but not under 532 nm (i.e. paramontroseite, VO_2). Visual evidence such as dark brown to black craters of the size of the beam spot or discoloration of the mineral surface are usually indicative of burning and dehydration, respectively.

Another not so often effect related to the heating of the sample by the incident laser is that while the spectra remains essentially the same, the peak positions shift to lower wavenumbers with increasing the energy of the laser. We detect this phenomenon when we compare the spectra from the 532 and 785 nm lasers and correct the lower wavenumber spectrum by lowering the intensity of the laser.

1.4.2 Effects of the incident laser wavelength

While the position of the peaks in the Raman spectrum remains mostly constant under different excitation wavelengths, their intensities can vary. As we have seen, we collect Raman scans using two different laser excitations from the same spot on the sample and in the same orientation. Therefore, the high-resolution narrow scans provide a

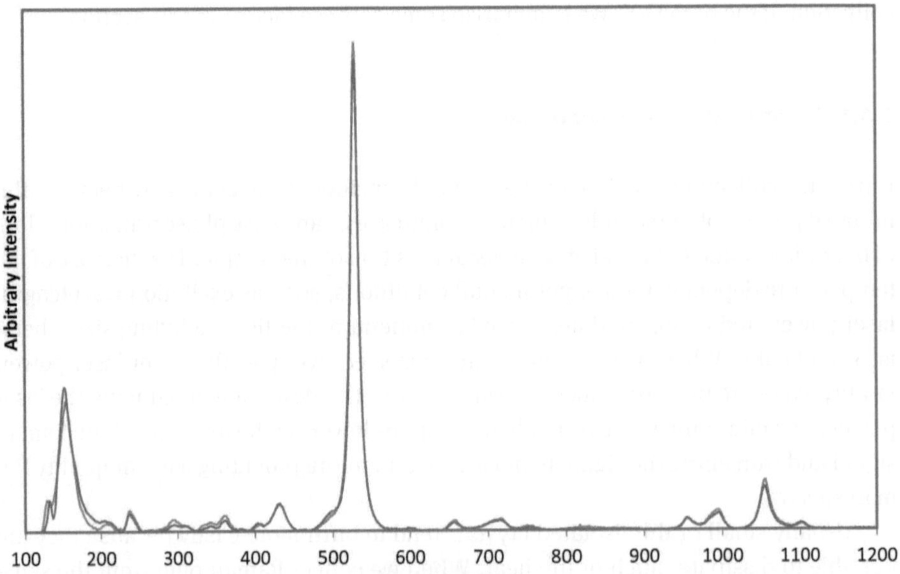

Fig. 1.11: High-resolution Raman spectra of the zeolite edingtonite (R040003), collected with both 532 (green line) and 785 (red line) nm lasers, illustrating an example of spectra whose patterns are essentially identical for both incident lasers. The sample was in identical position for both data collections, which eliminates the variation in intensities caused by orientation effects.

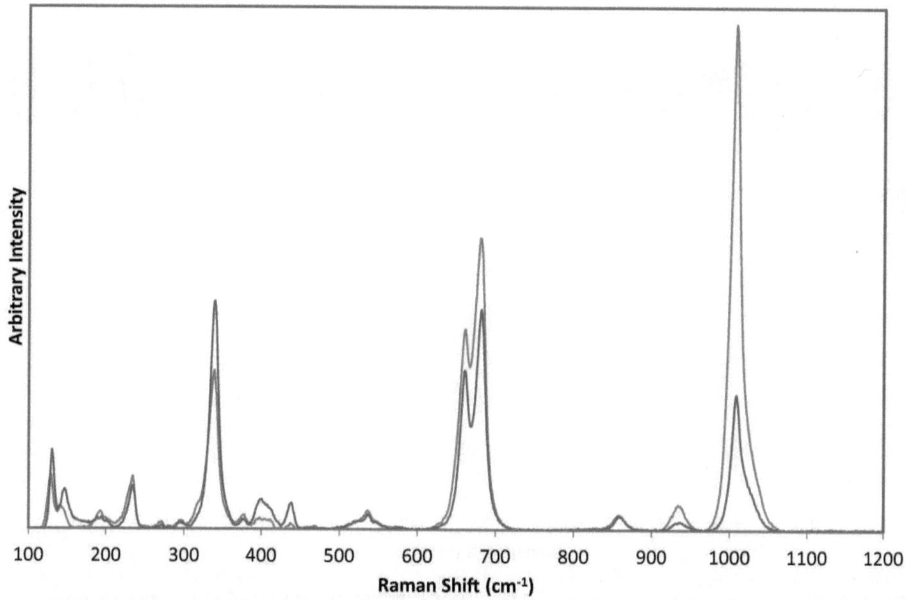

Fig. 1.12: High-resolution Raman spectra of enstatite (R040093), collected with both 532 (green line) and 785 (red line) nm lasers, illustrating an example of spectra whose patterns differ for both lasers. The sample was in identical position for both data collections, which eliminates the variation in intensities caused by orientation effects.

measure of how the spectra change with incident wavelength. We have seen that for some of the minerals, these two spectra may be the same or very alike, as in Fig. 1.11, but for others the intensities of identical modes can be different, as in Fig. 1.12.

1.4.3 Fluorescence

It has been found that about 16 % of the samples examined in the RRUFF project produced spectra that cannot be used for their identification, mostly due to the presence of broadband fluorescence or luminescence peaks dominating the Raman spectrum. Of these, half of them cannot be identified with either of the two lasers (532 and 785 nm) (Fig. 1.13), while the other half can be identified only with one of the lasers (Fig. 1.14).

Usually, moving to longer wavelengths of excitation can significantly reduce fluorescence [31] with the tradeoff of reducing the Raman signal because of the relation $I_{Raman} \sim 1/\lambda^4$ [25]. However, we have observed that some minerals fluoresce under 785 nm but not under 532 nm, as shown in Fig. 1.14.

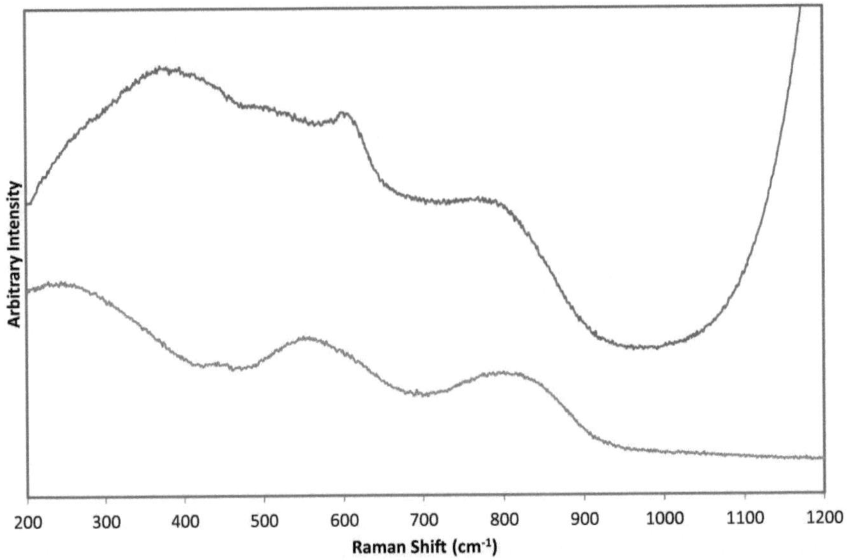

Fig. 1.13: High-resolution Raman spectra of loparite-Ce (R070251), collected with both 532 (green line) and 785 (red line) nm lasers, illustrating an example of spectra where both incident wavelengths are dominated by fluorescence.

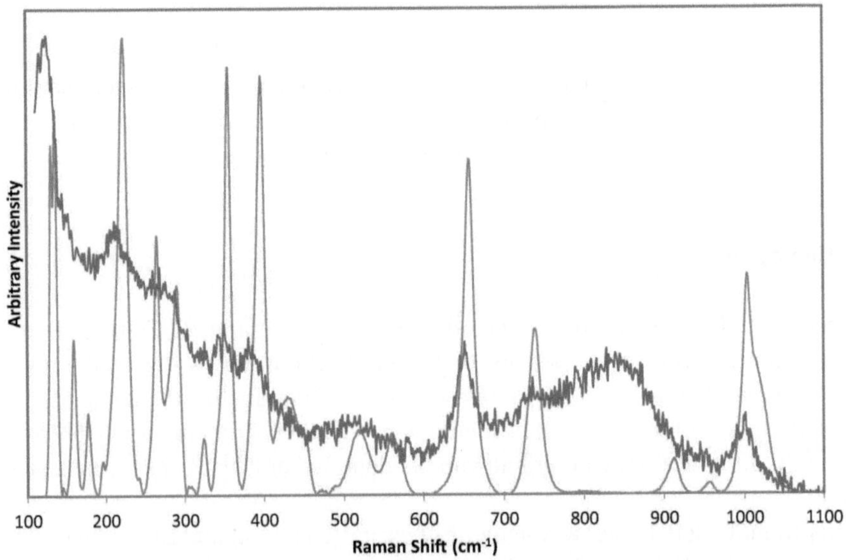

Fig. 1.14: High-resolution Raman spectra of dioptase (R040028), collected with both 532 (green line) and 785 (red line) nm lasers, illustrating an example of spectra where one of the incident wavelengths (780 nm laser) is associated with a poor pattern dominated by fluorescence.

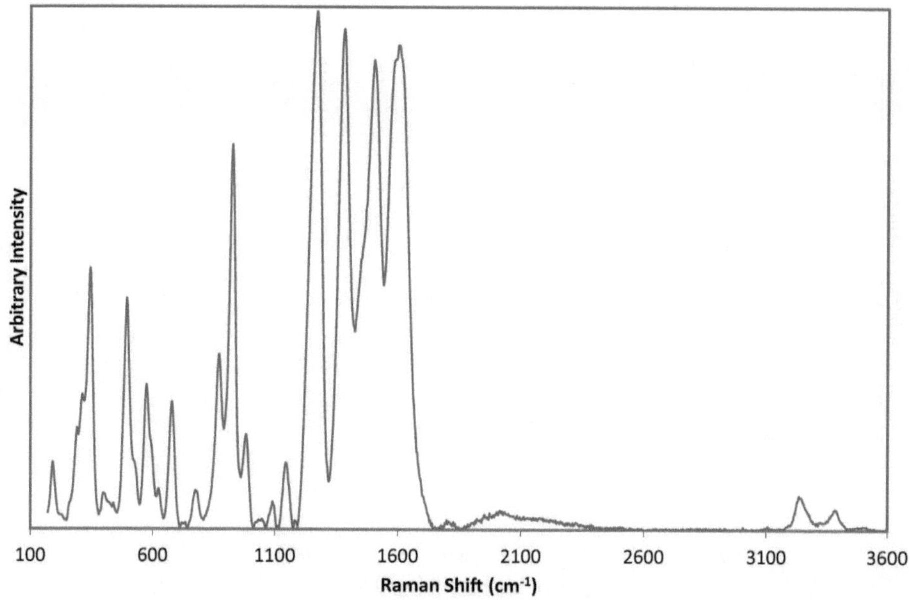

Fig. 1.15: Low-resolution Raman spectrum of zoisite (R060567), collected with the 785 nm laser, illustrating the significant luminescence features between 1200–1700 cm^{-1}. Such features were observed in the spectra of all the Ca-containing minerals from the deposits in Umba Valley, Tanzania [30]. The study proposes that these features are due to Nd^{3+} luminescence centers.

Some minerals may exhibit obvious luminescence peaks, as in Fig. 1.15, often due to the presence of transition metal or rare-earth ion centers [32], which in some situations, it can be used as a signature for the locality of the sample [30].

If the luminescence features are very intense, as in Fig. 1.16, they can obscure the Raman peaks of the particular mineral.

1.4.4 Presence of inversion center

Of the 4967 known mineral species, twenty-seven of them cannot produce a Raman signal because every atom is positioned on an inversion center. These include minerals that crystallize with the rocksalt (16) or copper (11) structures. In minerals containing an inversion center, the change in polarization of the atoms due to the incident beam is negated because the distortion of the electron cloud in one direction is equal to the distortion of the cloud in another direction. As a result, the Raman spectrum has intensity equal to zero. We have nevertheless collected the spectra for theses minerals and included them in the database (Fig. 1.17) because the absence of Raman activity also reveals important information about the sample. It is possible that some of these twenty-seven minerals may show weak Raman peaks. We interpret this as indicating

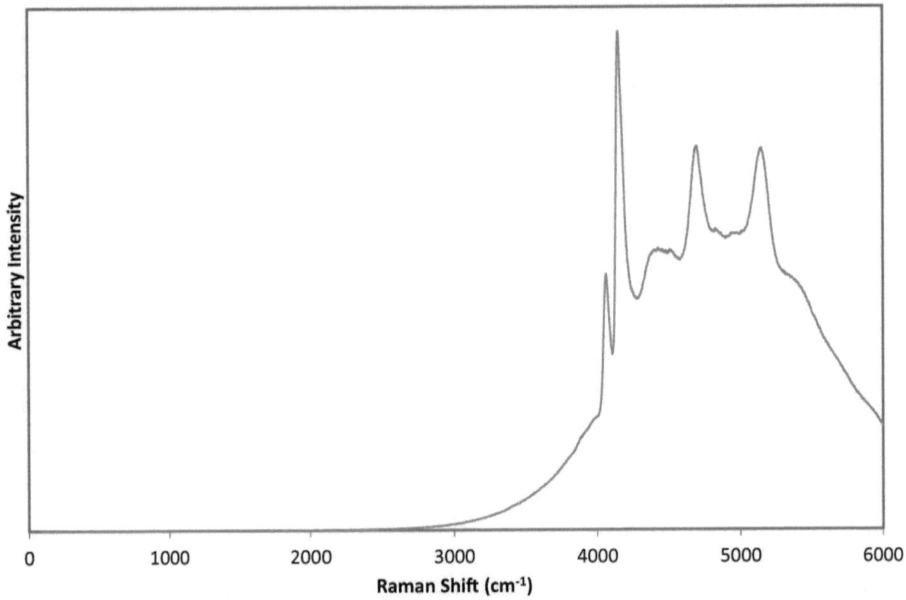

Fig. 1.16: Low-resolution Raman spectrum of unoriented topaz (R060024), collected with the 532 nm laser, illustrating intense luminescence features (4000–5200 cm^{-1}) that mask the Raman peaks under 1200 cm^{-1} (Fig. 1.10 shows the Raman bands of topaz in that specific region).

surface alteration. For instance, the oxidized surface of galena (PbS, rocksalt structure) may show weak peaks corresponding to minium (Pb_3O_4).

1.4.5 Metamict minerals

Metamict minerals are materials that were initially crystalline, but that have lost their crystallinity, partial or total, due to radioactive decay [33, 34]. Zircon [35], allanite [36], and titanite [37] are some examples of minerals that may exhibit metamictization. By annealing an unidentified metamict mineral at the appropriate temperature, its crystallinity can usually be restored and the sample identified by X-ray analysis. The RRUFF project studied eight metamict samples before and after heating to explore the effects on their Raman spectra. For example, the metamict gadolinite-(Y) (R060856) was identified and its Raman obtained after heating the sample at 1000ºC during 23 h [38].

Radiation damage produces change in the chemical and structural properties of minerals, and therefore, it can have dramatic effects on the Raman spectrum of a sample including loss of peak intensity, increase in peak width, and shifts in peak frequencies. For example, Nasdala et al. [39] calibrated the degree of metamictization in natural zircon crystals ($ZrSiO_4$) by measuring the broadening and shift towards lower

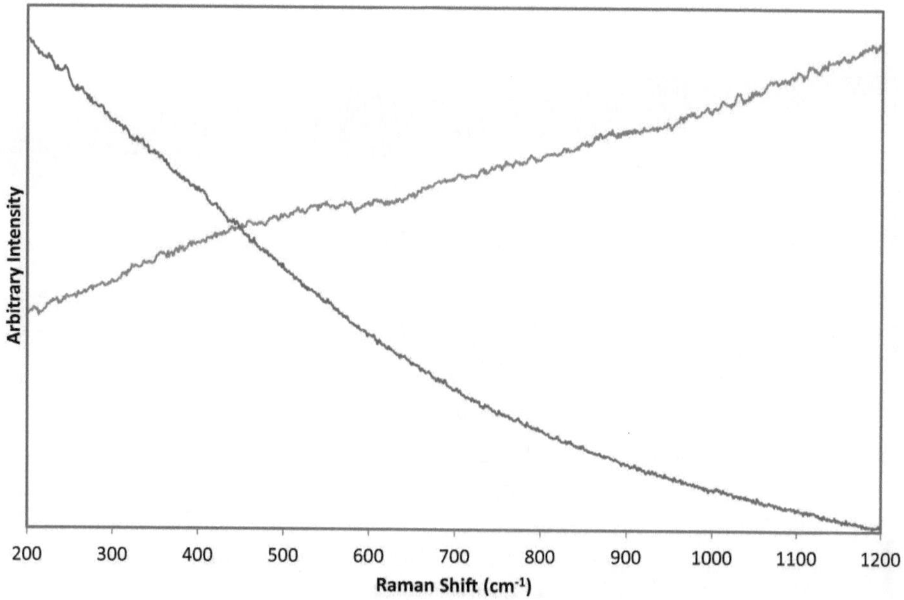

Fig. 1.17: High-resolution Raman spectra of halite (R070534), collected with both 532 (green line) and 785 (red line) nm lasers, illustrating the absence of Raman activity in minerals with every atom positioned on an inversion center.

wavenumbers of the Raman band at ~ 1008 cm^{-1} (ν_3SiO_4). Fig. 1.18 shows the variability in the position and width of the ν_3SiO_4 band in the Raman spectra of six zircon samples collected for the RRUFF project.

1.5 Applications of the RRUFF project

The power of the RRUFF project lies in its capability to aid in mineral identification and characterization. Records of well-characterized minerals can be used as a reliable Raman reference, and the ancillary software, including CrystalSleuth, XtalDraw, and XPOW, provides visualization and manipulation of data. Equally important to the RRUFF project is that its database should also provide the capability for inspiring other research efforts. For instance, a search on March 2015 of the Science Direct website (http://www.sciencedirect.com) retrieved a listing of 339 manuscripts that use data from the RRUFF project. These publications come from a broad variety of disciplines, including mineralogy, crystallography, geology, archeology, planetary science, art, medicine, material science, physics, chemistry, environmental, biology, and forensics.

The abundant and diverse collection of Raman spectra, X-ray diffraction patterns, and chemical analysis permit the systematic examination and analysis of mineral groups, chemical variations, phase transitions, etc. For example, using chemical anal-

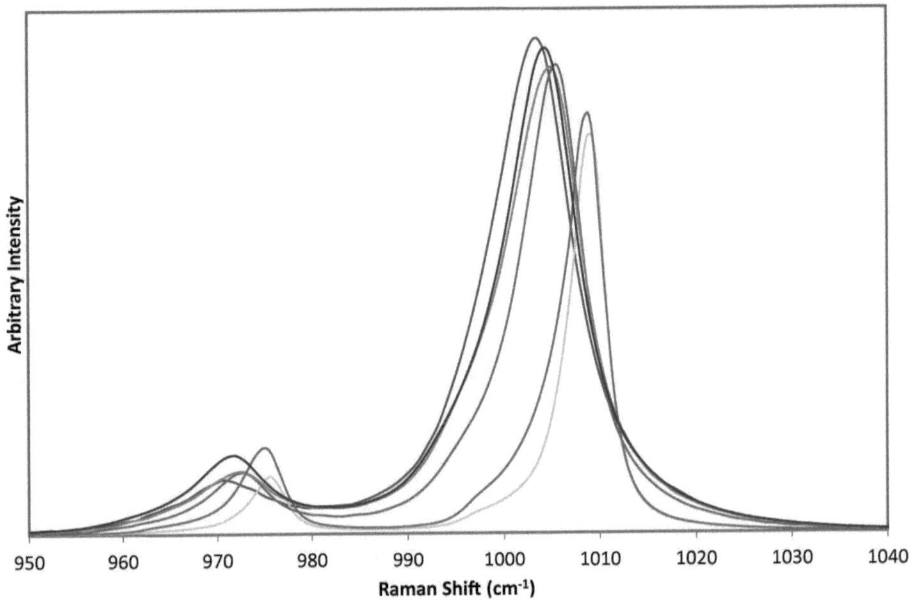

Fig. 1.18: Raman spectra in the 950–1040 cm^{-1} region of six unoriented zircons from the RRUFF database collected with 532 nm laser, showing the shift and broadening of the v_3SiO4 mode: R050034 (blue), R100144 (purple), R050488 (green), R050286 (aquamarine), R100128 (red), and R050203 (orange).

yses of 40 various garnets from the RRUFF project, Henderson and Downs [40] shows that the Raman spectra of silicate garnets are correlated to their empirical chemical compositions and can be used to estimate the compositions of unknown garnets to within 3 % atoms per formula unit, based on measurements of sixteen cations.

The collection of Raman spectra in the RRUFF database has had a profound impact on the gemology community. The inexpensive and non-destructive nature of Raman spectroscopy makes it a suitable technique for its use in this field, especially given the perfection of gemstones. The Raman spectrum of gemstones is not only used for their identification but also to determine their provenance (Fig. 1.15), as well as natural or synthetic origin and possible treatments applied to them [41]. For example, Lowry et al. [42] acquire Raman spectra from 96 semi-precious gemstones with the aim of classifying them with reference spectra from the RRUFF database. The RRUFF database and CrystalSleuth are used by Kuehn [43] to study the application of Raman spectroscopy for the identification of gemstones. The work by Jasinevicius [30] characterizes the vibrational and electronic features in the Raman spectra of many gemstones in the RRUFF database. The Raman instruments developed by M&A Gemological Instruments (MAGI) include the GemmoRaman software (http://www.gemmoraman.com/GemmoRaman-532.aspx) which provides a plugin for the automatic export of collected spectra into CrystalSleuth for their identification.

The variability in the RRUFF Raman spectra collection, reflected in chemical composition, laser wavelength, fluorescence, symmetry, site occupancy, and orientation, makes it suitable to use for building and testing mathematical models to automatically identify mineral phases from their Raman spectra. In Hermosilla Rodriguez et al. [44] they use one hundred randomly selected spectra from the RRUFF library to test an algorithm to identify mineral phases by Raman spectroscopy. In a recent publication, Cochrane and Blacksberg [45] develop a method to identify individual Raman features in mineral mixtures by using 11572 Raman spectra downloaded from the RRUFF project. The development of these models is of special importance for planetary science as they will be necessary in the analysis of the data coming from the Raman instruments planned for future space missions (see for example [46, 47]). A Raman spectrometer has been chosen as a part of the science instrument payload of the European Space Agency 2018 ExoMars mission to target mineralogical and astrobiological investigations on the surface and subsurface of Mars [48, 49] and NASA has just announced that its 2020 Mars Rover payload [50] will carry two Raman spectrometers to determine fine-scale mineralogy and detect organic compounds.

The identification and characterization of mineral specimens for the RRUFF project requires considerable research, mostly when something does not match with the available data. The lack of a match may occur because it is a new mineral, because its data was previously reported wrong, or because we do not have it yet in our database. So far, the IMA nomenclature commission has approved 21 new minerals discovered and described by our group (e.g. [51–54]) with other 50 or so anomalous specimens that require more careful examinations and currently remain non-public in the database.

We also study crystal structures of minerals with previously unreported structures (e.g. [55–58]), or minerals with problems on their previously reported structural data (e.g. [59, 60]). In our experience, the greatest problem with previously reported cell parameters from the older literature is found where twinning or subcell/supercell relations were not recognized, especially for data collected with point detectors. In some cases, the redefinition of the structure leads to a new ideal chemical formula for that mineral. For instance, esperite was first defined by Moore and Ribbe [61] as monoclinic with a well-developed "superlattice". However, Tait et al. [62] demonstrated that the so-called "superlattice" reflections were due to the presence of triple twins probably resulting from a high-to-low temperature phase transformation, which led them to the redefinition of the ideal chemical formula from $PbCa_3(ZnSiO_4)_4$ to $PbCa_2(ZnSiO_4)_3$. In the case of the redefinition of the vladimirite, the work by Yang et al. [63] demonstrates that while the ideal chemical formula of this mineral was correctly reported by Nefedov [64] as $Ca_4(AsO_4)_2(AsO_3OH){\cdot}4H_2O$, it had been modified incorrectly as $Ca_5H_2(AsO_4)_4{\cdot}5H_2O$ by subsequent studies [65].

1.5.1 Crystal-chemistry analysis of Mars samples

The RRUFF project is playing an important role in the study of the chemical composition of the mineral phases analyzed by the CheMin X-ray diffractometer [66] on the Mars Science laboratory rover Curiosity in Gale crater, Mars. Using the refined unit-cell parameters of samples analyzed at the sand dune, Rocknest [67, 68], Morrison et al. [69] estimates the empirical chemical formulas of the most abundant phases identified, including Mg-Fe-olivine, Mg-Fe-Ca-pyroxene and Na-Ca-K-feldspar. For the study, a large dataset of unit-cell parameters in combination with their associated chemical compositions were collected from the RRUFF project and the literature for each of these major phases. Moreover, by determining the chemical composition of the individual crystalline phases they were also able to estimate the bulk composition of the crystalline materials as well as the amorphous component. This methodology will be applied to each of the analysis performed by the CheMin instrument in the surface of Mars.

1.5.2 The Mineral Evolution database

The Mineral Evolution project [70] aims to understand how the episodes of planetary accretion and differentiation, plate tectonics, and origin of life lead to a selective evolution of mineral species through changes in temperature, pressure, and composition. In order to investigate questions related to these mineralogical changes through time and geological settings, the Mineral Evolution database [71] was created in partnership with the RRUFF project and Mindat.org (http://www.mindat.org). The goal of the database is to correlate the diversity of mineral species with their ages, localities, and other measurable properties. The RRUFF project serves as a platform to store and display the data accessed through the IMA-List. Mindat.org supplies the database with mineral localities and their associated mineral species.

1.6 Future directions

The RRUFF database was designed to include measurements restricted to microprobe analysis, Raman and infrared spectra, and X-ray diffraction patterns of minerals. This rigid structure prevents the addition of new measurements, such as Mössbauer, LIBS, or UV spectroscopy, without extensive and costly modification of the current software platform. The custom programmed solution used by RRUFF also makes it impracticable for other researchers to use with their data.

These limitations have led to the development of the Open Data Repository [72], an open-access platform for data publication funded by the Mars Science Laboratory CheMin team and NASA-Ames; developed by the University of Arizona's Department of Geosciences, and the Open Data Repository non-profit organization. The platform aims to provide researchers with a simple tool to publish any or all data related to their research on material properties. The data publication system makes it possible to share data publicly, to invite collaborators, and ensure research integrity.

With this system, users can create their own data repositories in their local server for their diverse data sets. Creating a data repository is a drag-and-drop procedure facilitated through a web-based forms designer. No direct interaction with the underlying database is required. The user can select and add different field types to their data sets and modify the design at any time to include new fields or file storage areas. The user can then start creating data records that describe individual pieces of data. A plugin system allows data sets containing XY data to be graphed and plotted to HTML output. Plug-ins for rendering chemical formulas and for displaying images in galleries are also bundled with the system.

The platform has been in development for over two years and is undergoing beta testing by a number of research groups. NASA's CheMin data from the Mars Science Laboratory is being published on the system as a proof of concept. Many new features and sharing tools will be developed under a new, 5-year contract with funding from NASA's SERA initiative and the Astrobiology Habitable Environments Database (AHED) [73] which will host data from a number of researchers in the astrobiology community.

In the long term, the migration of the RRUFF database into this new platform will make it possible to complement the current characterization of the mineral specimens with other techniques.

Acknowledgment: We are especially indebted to the support to the RRUFF Project by Mr. Michael Scott, who provided four million dollars in cash and 1000 rare mineral samples, as well as constant management advice and continuing encouragement. We also gratefully acknowledge support by the National Science Foundation EAR-0622371, Newmont Mining Corporation, Freeport-McMoRan Corporation, a Strategic University Research Partnership from the Jet Propulsion Laboratory, and the Arizona Science Foundation. This work is currently supported by the Changsha Mineral Show, NASA NNX11AP82A Mars Science Laboratory Investigations, and Science-Enabling Research Activity NASA-Ames. We thank the hundreds of mineral enthusiasts, both amateurs and professionals, who have contributed time and samples to the project. We also thank editors Rosa Micaela Danisi and Thomas Armbruster for their reviews and constructive comments.

Bibliography

[1] Dana JD. A system of mineralogy. New Haven, CT, USA, Durrie & Peck and Herrick & Noyes, 1837.

[2] Hintze C. Handbuch der Mineralogie. Leipzig, Germany, Verlag Von Veit & Comp, 1897–1933.

[3] Goldschmidt V. Altas der Krystallformen. Heidelberg, Germany, Carl Winters Universitätsbuchhandlung, 1913–1923.

[4] Vandenabeele P, Castro K, Hargreaves M, Moens L, Madariaga JM, Edwards HGM. Comparative study of mobile Raman instrumentation for art analysis. Anal Chim Acta 2007, 588, 108–16.

[5] Jehlička J, Vítek P, Edwards HGM, Hargreaves MD, Čapoun T. Fast detection of sulphate minerals (gypsum, anglesite, baryte) by a portable Raman spectrometer. J Raman Spectrosc 2009, 40, 1082–1086.

[6] Vítek P, Ali EMA, Edwards HGM, Jehlička J, Cox R, Page K. Evaluation of portable Raman spectrometer with 1064 nm excitation for geological and forensic applications. Spectrochim Acta A 2012, 86, 320–327.

[7] Kristova P, Hopkinson L, Rutt K, Hunter H, Cressey G. Quantitative analyses of powdered multi-minerallic carbonate aggregates using a portable Raman spectrometer. Am Mineral 2013, 98, 401–409.

[8] Henderson GS, Neuville DR, Downs RT. Spectroscopic methods in mineralogy and material sciences. Washington, DC, USA, The Mineralogical Society of America, 2014.

[9] McIntosh B, Denton MB, Downs RT, Becker D. Smart Raman instrument for Mars Science Laboratory. Mars Exploration Program Assessment Group, Sep 10–11, Jet Propulsion Laboratory, Pasadena, CA, USA, 2003.

[10] McIntosh B, Denton MB, Downs RT. Process Raman on Earth and research Raman on Mars – How are they related? Pittcon, March 11, Chicago, IL, USA, 2004.

[11] Downs RT. The RRUFF Project: an integrated study of the chemistry, crystallography, Raman and infrared spectroscopy of minerals. Program and Abstracts of the 19th General Meeting of the International Mineralogical Association in Kobe, Japan, 2006.

[12] Caracas R, Bobocioiu E. The WURM project – a freely available web-based repository of computed physical data for minerals. Am Mineral 2011, 96, 437–443.

[13] Anthony JW, Bideaux RA, Bladh KW, Nichols MC. Handbook of Mineralogy. Chantilly, VA, USA. Mineralogical Society of America, http://www.handbookofmineralogy.org/.

[14] Downs RT, Hall-Wallace M. The American Mineralogist Crystal Structure Database. Am Mineral 2003, 88, 247–250.

[15] Clark CM, Downs RT. Using the American Mineralogist Crystal Structure Database in the classroom. J Geos Educ 2004, 51, 76–80.

[16] Rajan H, Uchida H, Bryan DL, Swaminathan R, Downs RT, Hall-Wallace M. Building the American Mineralogist Crystal Structure Database: A recipe for construction of a small Internet database. In: Sinha A K, ed., Geological Society of America Special Paper 2006, 397, 73–80.

[17] Reed SJB. Electron Microprobe Analysis and Scanning Electron Microscopy in Geology, 2nd ed. New York, USA, Cambridge University Press, 2005.

[18] Deer WA, Howie RA, Zussman, J. An introduction to the rock-forming minerals, 3rd ed. London, UK, The Mineralogical Society, 2013.

[19] Lavina B, Dera P, Downs RT. Modern X-ray diffraction methods in mineralogy and geosciences. In: Henderson GS, Neuville DR, Downs RT eds. Washington, DC, USA, The Mineralogical Society of America, 2014, 1–31.

[20] Laetsch TA, Downs RT. Software for identification and refinement of cell parameters from powder diffraction data of minerals using the RRUFF Project and American Mineralogist Crys-

tal Structure Databases. Program and Abstracts of the 19th General Meeting of the International Mineralogical Association in Kobe, Japan, 2006.
[21] Downs RT, Bartelmehs KL, Gibbs GV, Boisen MB Jr. Interactive software for calculating and displaying X-ray or neutron powder diffractometer patterns of crystalline materials. Am Mineral 1993, 78, 1104–1107.
[22] Bartelmehs KL, Gibbs GV, Boisen MB Jr, Downs RT. Interactive computer software used in teaching and research in mineralogy at Virginia Tech. GSA Fall Meeting, Boston, MA, USA, 1993.
[23] Press WH, Teukolsky SA, Vetterling WT, Flannery BP. Numerical recipes in Fortran 77: the art of scientific computing, 2nd ed. New York, USA, Cambridge University Press, 1996.
[24] Yang H, Sano JL, Eichler C, Downs RT, Costin G. Iranite, $CuPb_{10}(CrO_4)_6(SiO_4)_2(OH)_2$, isomorphous with hemihedrite. Acta Crystallogr 2007, C63, i122–i124.
[25] Rull F. The Raman effect and the vibrational dynamics of molecules and crystalline solids. In: Dubessy J, Caumon MC, Rull F, eds. London, UK, EMU 2012, 1–60.
[26] Neuville DR, de Ligny D, Henderson GS. Advances in Raman spectroscopy applied to Earth and material sciences. In: Henderson GS, Neuville DR, Downs RT, eds. Washington, DC, USA, The Mineralogical Society of America 2014, 509–541.
[27] Griffith WP. Raman studies on rock-forming minerals. Part I. Orthosilicates and cyclosilicates. J Chem Soc (A) 1969, 1372–1377.
[28] Hill W, Rogalla D. Spike-correction of weak signals from charge-coupled devices and its application to Raman spectroscopy. Anal Chem 1992, 64, 2575–2579.
[29] Kroumova E, Aroyo MI, Perez-Mato JM, Kirov A, Capillas C, Ivantchev S, Wondratschek H. Bilbao Crystallographic Server: Useful databases and tools for phase-transition studies. Phase Transit 2003, 76, 155–170.
[30] Jasinevicius R. Characterization of vibrational and electronic features in the Raman spectra of gem minerals. Master Thesis, University of Arizona, Tucson, AZ, USA, 2009.
[31] Dubessy J, Caumon MC, Rull F, Sharma, S. Instrumentation in Raman spectroscopy: elementary theory and practice. In: Dubessy J, Caumon MC, Rull F, eds. London, UK, EMU, 2012, 83–172.
[32] Gaft M, Reisfeld R, Panczer G. Luminescence spectroscopy of minerals and materials, Heidelberg, Germany, Springer, 2005.
[33] Ewing RC. The crystal chemistry of complex niobium and tantalum oxides. IV. The metamict state: discussion. Am Mineral 1975, 60, 728–733.
[34] Ewing RC, Haaker RF. The metamict state: implications for radiation damage in crystalline waste forms. Nucl chem waste man 1980, 1, 51–57.
[35] Palenik CS, Nasdala L, Ewing RC. Radiation damage in zircon. Am Mineral 2003, 88, 770–781.
[36] Janeczek J, Eby RK. Annealing of radiation damage in allanite and gadolinite. Phys Chem Miner 1993, 19, 343–356.
[37] Hawthorne FC, Groat LA, Raudsepp M, et al. Alpha-decay damage in titanite. Am Mineral 1991, 76, 370–396.
[38] Lima de Faria J. Heat treatment of metamict euxenites, polymignites, yttrotantalites, samarskites, pyrochlores, and allanites. Mineralogical Museum, section of the Junta de Investigações do Ultramar 1958, 3R, 937–942.
[39] Nasdala L, Wenzel M, Vavra G, Irmer G, Wenzel T, Kober B. Metamictisation of natural zircon: accumulation versus thermal annealing of radioactivity-induced damage. Contrib Mineral Petrol 2001, 141, 125–144.
[40] Henderson RR, Downs RT. Determining chemical composition of the silicate garnets using Raman spectroscopy. Master Thesis, University of Arizona, Tucson, AZ, USA, 2009.

[41] Bersani D, Lottici PP. Applications of Raman spectroscopy to gemology. Anal Bioanal Chem 2010, 397, 2631–2646.
[42] Lowry S, Wieboldt D, Dalrymple D, Jasinevicius R, Downs RT. The use of a Raman spectral database of minerals for the rapid verification of semiprecious gemstones. Spectroscopy 2009, 24, 1–7.
[43] Kuehn JW. Raman and photoluminescence spectroscopy in gem identification. GIT proceedings Dec 8–9, Chiangmai, Thailand, 2014.
[44] Hermosilla Rodriguez I, Lopez-Reyes G, Llanos DR, Rull Perez F. Automatic Raman spectra processing for Exomars. In: Pardo-Igúzquiza E, Guardiola-Albert C, Heredia J, Moreno-Merino L, Durán JJ, Vargas-Guzmán JA, eds., Berlin, Germany, Springer 2014, 127–130.
[45] Cochrane CJ, Blacksberg J. A fast classification scheme in Raman spectroscopy for the identification of mineral mixtures using a large database with correlated predictors. IEEE Transactions on geoscience and remote sensing 2015, 53, 4259–4274
[46] Sobron P, Sobron F, Sanz A, Rull F. Raman signal processing software for automated identification of mineral phases and biosignatures on Mars. Appl Spectrosc 2008, 62, 364–370.
[47] Lopez-Reyes G, Sobron P, Lefebvre C, Rull F. Multivariate analysis of Raman spectra for the identification of sulfates: Implications for ExoMars. Am Mineral 2014, 99, 1570–1579.
[48] Rull F, Maurice S, Diaz E, Tato C, Pacros A, RLS Team. The Raman Laser Spectrometer (RLS) on the ExoMars 2018 Rover Mission. 42nd LPSC. Houston, TX, USA, 2011.
[49] Rull F, Sansano A, Díaz,E, et al. ExoMars Raman laser spectrometer for Exomars. Proc of SPIE 2011, 8152, 1–13.
[50] NASA Announces Mars 2020 Rover Payload to Explore the Red Planet as Never Before. NASA Press Release 14–208, Jul. 31, 2014. (Accessed April 8, 2015, at http://www.nasa.gov/press/2014/july/nasa-announces-mars-2020-rover-payload-to-explore-the-red-planet-as-never-before/).
[51] Yang H, Downs RT, Evans SH, Morrison SM, Schumer BN. Lefontite, IMA 2014–2075. CNMNC Newsletter No. 23, February 2015, page 55; Mineral Mag 2015, 79, 51–58.
[52] Yang H, Downs RT, Evans SH, Pinch WW. Lavinskyite, $K(LiCu)Cu_6(Si_4O_{11})_2(OH)_4$, isotypic with plancheite, a new mineral from the Wessels mine, Kalahari Manganese Fields, South Africa. Am Mineral 2014, 99, 525–530.
[53] Tait KT, Barkley MC, Thompson RM, Origlieri MJ, Evans SH, Prewitt CT, Yang H. Bobdownsite, a new mineral species from Big Fish River, Yukon, Canada, and its structural relationship with whitlockite-type compounds. Can Mineral 2011, 49, 1065–1078.
[54] Yang H, Jenkins RA, Thompson RM, Downs RT, Evans SH, Bloch EM. Markascherite, $Cu_3(MoO_4)(OH)_4$, a new mineral species polymorphic with szenicsite, from Copper Creek, Pinal County, Arizona, U.S.A. Am Mineral 2012, 97, 197–202.
[55] Lafuente B, Downs RT, Yang H, Jenkins RA. Calcioferrite with composition $(Ca_{3.94}Sr_{0.06})Mg_{1.01}(Fe_{2.93}Al_{1.07})(PO_4)_6(OH)_4{\bullet}12H2O$. Acta Cryst 2014, E70, i16–i17.
[56] Schumer BN, Downs RT, Domanik KJ, Andrade MB, Origlieri MJ. Pirquitasite, Ag_2ZnSnS_4. Acta Cryst 2013, E69, i8–i9.
[57] Origlieri MJ, Downs RT. Schaurteite, $Ca_3Ge(SO_4)_2(OH)_6{\bullet}3H_2O$. Acta Cryst 2013, E69, i6.
[58] Lafuente B, Yang H, Downs RT. Crystal structure of tetrawickmanite, $Mn^{2+}Sn^{4+}(OH)_6$. Acta Cryst 2015, E71, 234–237.
[59] Andrade MB, Doell D, Downs RT, Yang H. Redetermination of katayamalite, $KLi_3Ca_7Ti_2(SiO_3)_{12}(OH)_2$. Acta Cryst 2013, E69, i41.
[60] Morrison SM, Downs RT, Yang H. Redetermination of kovdorskite, $Mg_2PO_4(OH){\bullet}3H_2O$. Acta Cryst 2012, E68, i12–i13.
[61] Moore PB, Ribbe PH. A study of "calcium-larsenite" renamed esperite. Am Mineral 1965, 50, 1170–1178.

[62] Tait KT, Yang H, Downs RT, Li C, Pinch WW. The crystal structure of esperite, with a revised chemical formula, $PbCa_2(ZnSiO_4)_3$, isostructural with beryllonite. Am Mineral 2010, 95, 699–705.
[63] Yang H, Evans SH, Downs RT, Jenkins RA. The crystal structure of vladimirite, with a revised chemical formula, $Ca_4(AsO_4)_2(AsO_3OH) \bullet 4H_2O$. Can Mineral 2011, 49, 1055–1064.
[64] Nefedov EI. Report on new minerals discovered by him. Zap Vses Mineral Obshchest 1953, 82, 311–317 (in Russian).
[65] Pierrot R. Contribution à la minéralogie des arséniates calciques et calcomagnésiens naturels. B Soc Fr Mineral Cristallogr 1964, 87, 169–211 (in French).
[66] Blake D, Vaniman D, Achilles C, et al. Characterization and calibration of the CheMin mineralogical instrument on Mars Science Laboratory. Space Sci Rev 2012, 170, 341–399.
[67] Blake DF, Morris RV, Kocurek G, et al. Curiosity at Gale Crater, Mars: Characterization and analysis of the Rocknest sand shadow. Science 2013, 341, 1239505, 1–7.
[68] Bish DL, Blake DF, Vaniman DT, et al. X-ray diffraction results from Mars Science Laboratory: Mineralogy of Rocknest at Gale Crater. Science 2013, 341, 1238932, 1–5.
[69] Morrison SM, Downs RT, Blake DF, et al. Crystal-chemical analyses of Martian minerals in Gale crater. 21st General Meeting of the IMA, Gauteng, South Africa, 2014.
[70] Hazen RM, Papineau D, Bleeker W, et al. Mineral evolution. Am Mineral 2008, 93, 1693–1720.
[71] Hazen RM, Bekker A, Bish DL et al. Needs and opportunities in mineral evolution research. Am Mineral 2011, 96, 953–963.
[72] Stone N, Lafuente B, Downs RT, Blake D, Bristow T, Fonda M. The Open Data Repository's Data Publisher. AbSciCon, Chicago, IL, USA, 2015.
[73] Lafuente B, Stone N, Downs RT, Blake D, Bristow T, Fonda M. The Astrobiology Habitable Environments Database (AHED). AbSciCon, Chicago, IL, USA, 2015.

Sergey V. Krivovichev

2 Structural complexity of minerals and mineral parageneses: Information and its evolution in the mineral world

2.1 Introduction

One of the important aspects of mineralogy as a sub-discipline of geology is its bearing on the chemical composition and structural features of minerals and their evolution in space and time. Thus mineralogy is both a nomothetic and idiographic discipline [1]. As a nomothetic science, it establishes general laws that govern particular features of minerals as crystalline materials of natural origin, such as relations between structure and chemical composition, spectroscopic properties, response to changing thermodynamic conditions, etc. As an idiographic discipline, mineralogy is oriented towards the past and tries to reconstruct particular conditions of mineral formation on different scales, from local nanometer interactions to the global evolution of matter in the Universe. In the framework of the current chemical [2] and mineralogical [3] thought, minerals evolve together with the other matter in the Universe and reconstruction of this evolution is one of the main tasks of mineralogy. Back in the 1980s, N.P. Yushkin [4] realized the need in uniform quantitative parameters that would reflect such features of natural crystalline matter as symmetry, chemical and structural diversity, and complexity. The advent of the modern theory of complexity and its increasing applications in various fields of science [5–8], as well as intensive accumulation and cataloging of mineralogical data [9], provides an excellent framework for the development of detailed theories of diversity and complexity of minerals and analysis of their evolution in the history of the Earth and the observable Universe. This chapter touches upon only one aspect of mineral complexity, that is, the complexity of crystal structures of minerals. The fact that mineral structures may have different degrees of complexity has long been recognized, which is reflected in the titles of scientific papers such as 'Serendibite, a complicated, new, inorganic crystal structure' (Buerger, Venkatakrishnan [10]), ' Crystal structure of steenstrupine: A rod structure of unusual complexity' (Moore, Shen [11]), 'Structure and hydrogen bonding in preobrazhenskite, a complex heteropolyhedral borate' (Burns, Hawthorne [12]), 'The crystal structure of dalnegroite: A masterpiece of structural complexity' (Bindi et al. [13]), etc. However, only recently the very versatile concept of complexity received its solid definitions in terms of Shannon information theory [14–16]. The present chapter provides an overview of the theoretical framework to describe and to interpret structural complexity of minerals and is organized as follows. The second section is devoted to the different definitions of complexity, of which the most emphasis is given to the information-based parameters, statistics of which is considered in the third section. The fourth part deals with relations between complexity and hierarchical features of structural organization, whereas the fifth, sixth, seventh and eighth sections are devoted to the applications of information-based complexity parameters to the interpretation of micro- and macroscale mineral-forming processes. The nineth section outlines major differences between minerals and living matter, in terms of both its complexities and diversities. The last section provides the main conclusions and directions for further research.

2.2 Measuring structural complexity

2.2.1 Parsimony and its indices

Probably, the first consideration of structural complexity, at least in the published literature, should be assigned to Linus Pauling's paper on 'The principles determining the structure of complex ionic crystals' [17]. Among his five rules, the fifth rule of parsimony deals directly with structural complexity. In its original formulation, it says that 'the number of different types of constituents in a crystal tends to be small.' Pauling [17] did not provide any quantitative procedure to evaluate structural parsimony, and this was done by Baur et al. [18], who proposed two types of **parsimony indices** in order to describe topological and geometrical complexity of crystal structures.

The topological parsimony index, I_t, was defined by Baur et al. [18] as

$$I_t = (t - e)/t, \tag{2.1}$$

where t is the number of topologically distinct atom sites in the asymmetric unit, and e is the number of different chemical elements.

The crystallographic parsimony index, I_c, can be calculated as

$$I_c = (c - e)/c, \tag{2.2}$$

where c is the number of crystallographically distinct atom sites. Within this approach, topological and crystallographic equivalence of atoms are understood in terms of topological and crystallographic symmetry groups, respectively. Topological symmetry group of an atomic arrangement is the highest symmetry group that the arrangement may adopt by continuous transformation that does not involve breaking of any bonds between atoms. Crystallographic symmetry corresponds to the space group of the structure. Topological symmetry may be either higher or identical to (but not lower than) the crystallographic symmetry. This distinction between crystallographic and topological symmetry will be used in the following throughout the paper.

Baur et al. [18] pointed out that the topological parsimony index is of primary importance and suggested subdivision of inorganic structures into lavish (I_t >0.66), intermediate ($0.66 < I_t < 0.33$) and parsimonius ($I_t < 0.33$). However, this index has little discriminative power: for instance, $I_t = I_c = 0$ for cubic halite (NaCl), tetragonal rutile (TiO_2), and cubic $BaTiO_3$. Both indices are not sensitive to chemical complexity or to symmetry and reflect only relations between the number of chemical elements and the number of topologically distinct atom sites. Though parsimony is generally related to structural complexity, the parsimony indices proposed by Baur et al. [18] cannot be used as unambiguous measures of structural complexity of crystals.

Development of new methods of crystal-structure prediction and simulation led Oganov and Valle [19] to the formulation of a fingerprint function, which is a function that describes a crystal structure in a unique way on the basis of a specified geometrical environment of each atom site. In the framework of this approach, each structure can be visualized as a vector in the abstract fingerprint function-space (see [19] for a detailed definition). In the same manner, each site in a crystal structure has its own fingerprint function, depending upon its particular geometrical surrounding within a sphere with the specified radius R. The difference between the two structures (or two atom sites) is the distance D between their fingerprints in a fingerprint space. In order to quantify complexity of a particular crystal structure, Oganov and Valle [19] introduce quasi-entropy, S_{str}, defined as

$$S_{str} = \Sigma_A (N_A/N)\langle ln(1 - D_{AiAj})\rangle, \tag{2.3}$$

where A_i and A_j are the ith and jth sites occupied by the same chemical element A, N_A is the number of atoms of this element in the unit cell, N is the total number of atoms in the unit cell, and D_{AiAj} is the distance between the fingerprint functions of A_i and A_j. If a particular element occupied only one site in a crystal structure, $D_{AiAj} = 0$ and therefore $S_{str} = 0$.

On the basis of energy calculations for a large number of potential crystal structures with the same composition and comparison of the obtained energies with their structural quasi-entropies, Oganov and Valle [19] conclude that '...the ground state normally adopts one of the simplest structures compatible with the chemistry of the compound. Such structures tend to have lower energies. This rule is nothing else than an energetic-based reformulation of Pauling's fifth rule.' This also means that '...in the ground state and low-energy structures, atoms of each species tend to occupy similar crystallographic sites.' The work by Oganov and Valle (2009) is probably the most spectacular development of the Pauling's line of thought on the complexity of crystal structures, since it provides a unique way of correlation of complexity with ground-state energies. However, despite of its enormous importance, this approach suffers the same disadvantage as the parsimony indices proposed by Baur et al. [18]: if each chemical element occupies a unique crystallographic site in the structure, its quasi-entropy is equal to zero, irrelevant of its symmetry and the number of different chemical elements (chemical complexity). Thus, $S_{str} = 0$ for diamond (C), halite (NaCl), rutile (TiO_2) and cubic perovskite ($BaTiO_3$).

An alternative approach to structural complexity was suggested by Lister et al. [20, 21], who used the number of atoms in the asymmetric unit as a measure of complexity for a commensurate structure. This method has many advantages for the comparison of crystal structures of low symmetry, but, unfortunately, does not provide appropriate quantification of complexity of high-symmetry structures such as, e.g. paulingite, a complex natural zeolite, which was identified by Mackay and Klinowski [22] as one of the most complex inorganic structures.

The disadvantages of the parsimony indices in defining structural complexity of crystals originated from the fact that they are purely crystallographic. In order to define such a category as crystal-structure complexity, one has to quit the world of crystallographic parameters and turn to the modern definitions of complexity that have their origin in the field of information theory and computer science.

2.2.2 Information-based complexity measures

In the framework of the complexity theory, one of the approaches to the complexity of an object is to measure the amount of information that it contains: **the more information the object contains, the more complex it is**. In order to apply this principle to the crystal structures of minerals, we first provide an overview of the application of information theory to spatial structures.

2.2.2.1 Shannon formula and information content of a message

The concept of information was developed in connection with the theory of communication [23]. Information is measured in bits, or binary digits. Suppose that we have a message M consisting of n symbols taken from an alphabet $\mathbf{S}$ consisting of precisely s different symbols. The probability p_i of the appearance of the ith symbol from $\mathbf{S}$ is equal to

$$p_i = n_i/n, \tag{2.4}$$

where n_i is the number of the ith symbols in M.

Therefore

$$\sum_{i=1}^{s} p_i = 1. \tag{2.5}$$

The information content of the message M is calculated according to the *Shannon formula*

$$H = -\sum_{i=1}^{s} p_i \log_2 p_i \quad \text{(bits/symbol)}. \tag{2.6}$$

In the case of a single coin toss, there are two possible outcomes: head (0) or tail (1), and the amount of information is equal to 1 bit. In the case of two sequential coin tosses, there are four equally probable outcomes: 00, 01, 10, and 11, and the amount of information is equal to 2 bits. If a process has three equally probable outcomes, the amount of information registered by the system is $H = \log_2 3 = 1.585$ bits.

2.2.2.2 Information content and complexity of a graph

A graph, G, is defined as a structure consisting of a set of objects $V = \{v_1, v_2, \ldots\}$t called vertices, and a set of unordered pairs of vertices $E = \{v_i, v_j\}$, whose elements are called edges. Graphs have found extensive application in crystal chemistry and structural mineralogy, e.g. when atoms (or coordination polyhedra) are identified with vertices and bonds (or interpolyhedral links) with edges [24, 25].

Rashevsky [26] was the first to apply the Shannon formula to the information content of topological and geometrical structures. He considered a graph as a message and its vertices as symbols. Two vertices of a graph G are equivalent if their permutation does not result in a graph topologically different from (= non-isomorphic to) G. For instance, the graph shown in Fig. 2.1a consists of five vertices, which can be subdivided into three equivalence classes, C_1, C_2 and C_3. The first class, C_1, consists of vertices 3 and 4 ($C_1 = \{3, 4\}$); these vertices are equivalent as their permutation does not change the topology of the graph. The same is true for vertices 2 and 5, which belong to the class C_2. The third class, C_3, consists of the vertex 1, which has no equivalent vertex in the graph. The probability of a vertex from a certain class to be chosen randomly is equal to 2/5 for C_1 and C_2, and 1/5 for C_3: $p_1 = p_2 = 0.4$ and $p_3 = 0.2$. According to Rashevsky [26] and Trucco [27], the structural information content, I_G, of a graph, G, is defined as

$$I_G = -\sum_{i=1}^{k} p_i \log_2 p_i \quad \text{(bits/vertex)}, \qquad (2.7)$$

where p_i is the probability of occurrence of the vertex of the ith class.

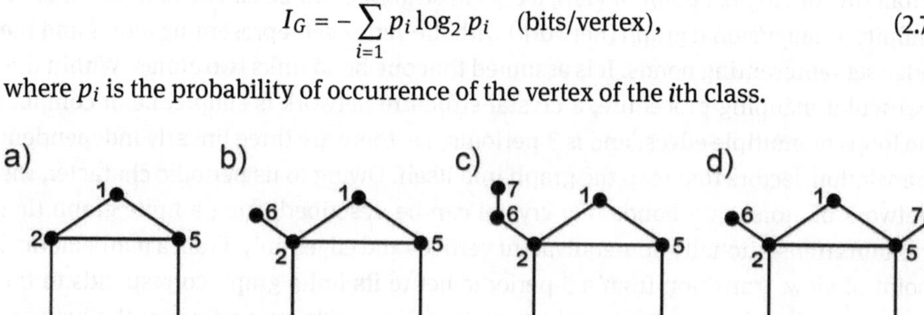

Fig. 2.1: Graphs of different complexity. See text for details. Reproduced with the kind permission of the Mineralogical Society from [15], published in Mineralogical Magazine (2013).

The information content of the graph shown in Fig. 2.1a is

$$I_G 0 - [0.2 \log_2(0.2) + 0.4 \log_2(0.4) + 0.4 \log_2(0.4)] = 1.53 \text{ bits/vertex}.$$

Its total information content, $I_{G,\text{total}}$, is equal to $5 \times 1.53 = 7.65$ bits.

The graphs shown in Fig. 2.1b,c,d illustrate the evolution of structural information content along the topological modifications of the initial graph shown in Fig. 2.1a. Addition of an additional vertex, (6), and an additional edge (2)–(6) to the vertex 2 (Fig. 2.1b) results in a dramatic increase in I_G, from 1.53 to 2.62 bits/vertex (the total information content increases from 7.65 to 15.72 bits, i.e. more than twice!). The reason for such an increase in information is that addition of vertex 6 and edge 2–6 destroys the symmetry of the original graph; now all the vertices of the graph are inequivalent. Addition of vertex 7 and edge 6–7 (Fig. 2.1c) results in a slight increase in I_G, from 2.62 to 2.80 bits/vertex ($I_{G,\text{total}}$ changes from 15.72 to 19.60 bits). However, addition of vertex 7 and edge 5–7 to the graph shown in Fig. 2.1b restores the symmetry of the original graph (Fig. 2.1d), and the information content drops from 2.62 to 1.96 bits/vertex. The total information content changes from 15.72 to 13.72 bits, i.e. the information content of the graph with 6 vertices shown in Fig. 2.1b is higher than that for the graph with 7 vertices shown in Fig. 2.1d.

The seminal work by Rashevsky [26] was followed by extensive work summarized by Dehmer and Mowshowitz [28]. The information content of a graph has been considered as measure of its complexity: the more information the graph contains, the more complex it is. A number of different and complicated information-based complexity measures have been developed for complex graphs and networks of different nature and origin [6].

2.2.2.3 Crystal structure as a message

From the topological point of view, a crystal structure can be can be mapped onto an infinite 3-dimensional graph (network) with the vertex set representing atoms and the edge set representing bonds. It is assumed that one bond links two atoms. Within this particular mapping procedure, a crystal-structure network is simple, i.e. it contains no loops or multiple edges, and is 3-periodic, i.e. there are three linearly independent translation vectors that map the graph into itself. Owing to its periodic character, the network of atoms and bonds in a crystal can be described using a finite graph that contains translationally non-equivalent vertices and edges only. From a mathematical point of view, transition from a 3-periodic net to its finite graph corresponds to the operation of projection, where all translationally equivalent vertices of the network project onto one vertex of the finite graph. This graph is called a *quotient graph* of a 3-periodic net [29, 30].

Fig. 2.2a shows a simple 2-periodic graph consisting of black and white vertices. Its quotient graph (Fig. 2.2b) is quite simple, having 3-connected black and 2- and 4-connected white vertices. Fig. 2.3a shows the structure of lopezite, $K_2Cr_2O_7$, a rare soluble secondary mineral discovered by Bandy [31] in vugs in massive nitrate rocks in the Tarapaca mine, Chile. The quotient graph of the structure is shown in Fig. 2.3b, and one can easily recognize its complexity.

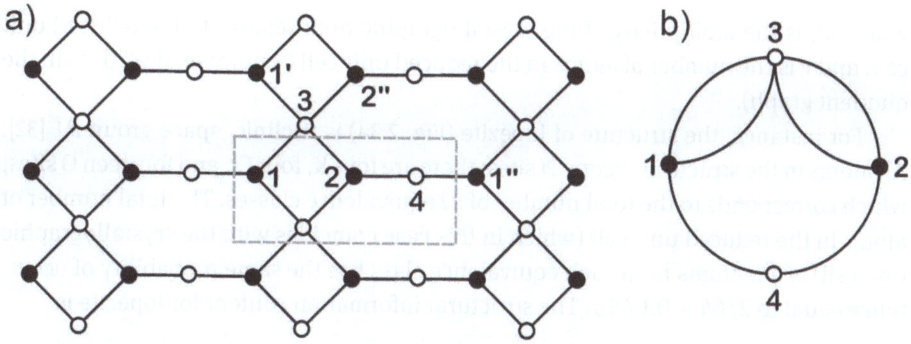

Fig. 2.2: Two-dimensional periodic network (a) and its quotient graph (b). Reproduced with the kind permission of the Mineralogical Society from [15], published in Mineralogical Magazine (2013).

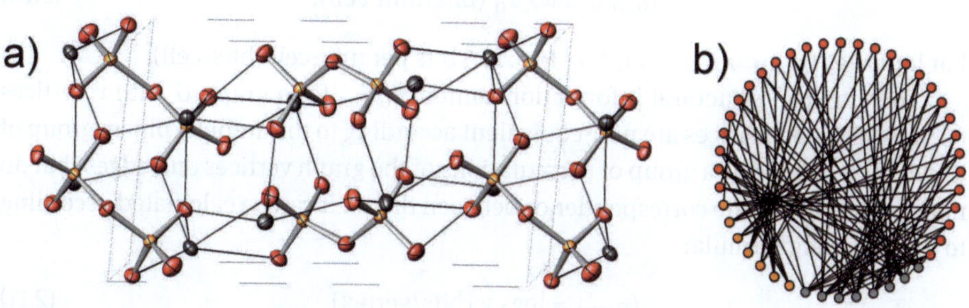

Fig. 2.3: The crystal structure of lopezite, $K_2Cr_2O_7$ (a), and its quotient graph (b). Reproduced with the kind permission of the Mineralogical Society from [15], published in Mineralogical Magazine (2013).

As the quotient graph completely represents crystal-structure topology, its complexity has a direct correspondence to the complexity of the whole structure. Application of the Rashevsky formula [eq. (2.7)] requires equivalence criteria to be defined for the graph vertices. Krivovichev [14] suggested subdividing the vertices of a quotient graph into equivalence classes according to the Wyckoff sites that corresponding atoms occupy in the crystal structure. The important condition is that the multiplicity of a site should be related to the *reduced* (i.e. primitive) unit-cell.

The following formula was proposed to calculate the *structural information content* of a crystal structure [14]:

$$I_G = -\sum_{i=1}^{k} p_i \log_2 p_i \quad \text{(bits/atom)}, \tag{2.8}$$

where k is the number of crystallographic orbits and p_i is the probability of occurrence of the atom of the ith crystallographic orbit, that is:

$$p_i = m_i/v, \tag{2.9}$$

where m_i is the multiplicity of the crystallographic orbit relative to the reduced unit cell, and v is the number of atoms in the reduced unit-cell (= number of vertices in the quotient graph).

For instance, the structure of lopezite (Fig. 2.3a) is triclinic, space group $P\bar{1}$ [32]. All atoms in the structure occupy 2*i* sites: there are four K, four Cr, and fourteen O sites, which corresponds to the total number of 22 equivalence classes. The total number of atoms in the reduced unit-cell (which in this case coincides with the crystallographic unit cell) is 44. Atoms from each equivalence class has the same probability of occurrence equal to $2/44 = 0.0455$. The structural information content for lopezite is

$$I_G = -22 \times (2/44) \log_2 2/44 = 4.459 \text{ bits/atom}.$$

The total information content is determined by the formula

$$I_{G,\text{total}} = v \times I_G \text{ (bits/unit-cell)}. \tag{2.10}$$

For lopezite, $I_{G,\text{total}} = 44 \times 4.459 = 196.196$ bits per unit-cell (bits/cell).

The maximal structural information content $I_{G\,\text{max}}$ for a graph G with v vertices occurs, where all vertices are non-equivalent according to the automorphism group of G (which is defined as a group of permutations of the graph vertices and edges that do not change one-to-one correspondence between them). It can be calculated according to the following formula:

$$I_{G\,\text{max}} = \log_2 v \text{ (bits/vertex)}. \tag{2.11}$$

In order for I_G to be independent of v, we may define the normalized structural information content, $I_{G,\text{norm}}$, as

$$I_{G,\text{norm}} = I_G/I_{G\,\text{max}} = -\left[\sum_{i=1}^{k} p_i \log_2 p_i\right] / \log_2 v. \tag{2.12}$$

For the structure of lopezite, $I_{G\,\text{max}} = \log_2 44 = 5.459$ bits/atom and $I_{G,\text{norm}} = 4.459/5.459 = 0.817$.

The approach described above considers structure as a message consisting of atoms; the complexity of the structure is determined by its information content calculated through the direct application of the modified Shannon formula. The I_G and $I_{G,\text{norm}}$ parameters are sensitive to the symmetry of the structure and thus describe its combinatorial complexity, whereas $I_{G,\text{total}}$ combines both symmetry and size of the structure, and therefore combines combinatorial and symbolic complexities in the context discussed by Steurer [33].

2.2.2.4 Information density

The information-based complexity measures allow us to introduce the concept of *information density*, which can be defined as a number of bits of information per cubic ångström:

$$\rho_{\text{inf}} = I_{G,\text{total}}/V\ (\text{bits}/\text{Å}^3). \tag{2.13}$$

For lopezite, $\rho_{\text{inf}} = 0.270$ bits/Å^3. For another modification of $K_2Cr_2O_7$ [34], $\rho_{\text{inf}} = 0.156$ bits/Å^3, which indicates its lower complexity compared to lopezite. It is important to note that information density should be calculated using the volume of the reduced unit-cell.

For two structures with similar unit cells but different symmetry, information density will be higher for the structure with lower symmetry (and higher complexity). For instance, cation ordering within the same unit cell would lead to the formation of superstructure with higher information density than that of the original structure. This shows that unit-cell volume alone cannot be considered as a measure of structural complexity, since it reflects only symbolic complexity of the system and is not sensitive to its combinatorial complexity.

Information density can be used also as a measure of complexity for crystal structures of similar composition. The fact that complexity is a function of both information content and unit-cell volume makes it dependent upon the size of atoms and ions that constitute the respective crystal, as well as upon temperature and pressure. Suanite, $Mg_2B_2O_5$ [35], and natrosilite, $Na_2Si_2O_5$ [36], have the same total structural information content (114.117 bits/cell), but their information densities are drastically different (0.333 and 0.242 bits/Å^3, respectively), which is an obvious consequence of the different ionic radii of the Mg^{2+} and Na^+ ions, on one hand, and of the B^{3+} and Si^{4+} ions, on the other.

2.2.3 Algorithmic complexity

2.2.3.1 General notes and historical remarks

The complexity measures based upon Shannon information is just one class of complexity parameters known in the modern 'science of complexity' [37]. In fact, Shannon information describes the **static** complexity, which addresses purely structural informational aspects and is independent of the processes by which information is encoded and decoded. Another type of complexity measures describes **dynamic** complexity, which addresses the question of how much dynamical or computational effort is required to build a system [38]. Among the dynamic complexity measures, one of the most popular is algorithmic complexity (or Kolmogorov complexity) that is defined as the length of the shortest program such as, when executed, yields the particular state of a system.

Estevez–Rams and González–Férez [39] discussed the use of algorithmic complexity to define the notion of crystallinity. They proposed to calculate the Kolmogorov complexity $K(s/N)$ of the structure s consisting of N atoms according to the following formula:

$$K(s/N) = /r^*/ + /v/ + /a^*/, \tag{2.14}$$

where r^* is the set of generators of the space group, v^* are atom coordinates, and a^* is the algorithm to compute all atomic positions from r^* and v^*. The notion $/s/$ means 'the length of the string s'. In this definition, $K(s/N)$ provides a universal measure of the process of generating crystal structures from lattice parameters, space-group generators and atomic coordinates. The measure can be applied to any crystal structure without regard to chemical composition and topology. However, there is no clear procedure to calculate the $K(s/N)$ value; in particular, Estevez-Rams and González-Férez [39] provided no example of such a calculation, which makes this parameter non-applicable in its present form. Moreover, the $K(s/N)$ parameter would provide a numerical description of the theoretical construction of a crystal structure (as implemented, for instance, in structure-drawing software) that has no relation to the process of formation of a crystal. A more natural and reality-relevant way to measure algorithmic complexity of a crystal structure is to investigate possible processes of building structure by attachment of atomic particles to the surface of a growing crystal. In the science of crystal growth, one of the accepted and widely used is the Kossel-Stransky or Terrace-Ledge-Kink (TLK) model, where crystal growth is viewed as happening due to the sequential attachment of cubic particles to the surface of a crystal. This model as well as the model of growth on dislocations was useful to describe some thermodynamic and kinetic aspects of crystal growth, but it is far from the real world in that the real particles are not ideal cubes and crystal-structure topology is far from the ideal cubic lattice. More realistic models can be developed by the analysis of topologies of real structures and the way of their generation by preferential attachment of atomic particles or building blocks. Mackay [40] pointed out that crystal growth can be modeled using finite automata and, in particular, cellular automata. This idea was applied by Krivovichev [41] to modeling growth of structures consisting of cubic modules such as sulfides of the djerfisherite and bartonite groups. The cellular automata models have also been used to generate novel framework types in zeolites [42], to describe growth and disorder of heteropolyhedral layered structures in uranyl selenates [43, 44], and to discuss possible pathways of self-assembly in the lovozerite-type structures [45]. It has been demonstrated that different periodic topologies may be generated by the same automaton depending upon the initial conditions (structure of a 'nuclei').

Krivovichev et al. [46] suggested using finite automata to describe generation of orthogonal networks and to use complexity of the state diagrams to evaluate algorithmic complexity of resulting structures. This method was applied to a wide range of inorganic structures, including minerals, zeolites, inorganic oxysalts, and coordination polymers. Krivovichev [47] demonstrated that a large amount of known topologies can be described as orthogonal networks, which allows construction of their topology-generating finite automata and further analysis of their complexity. Below we provide an overview of the use of finite automata to define algorithmic notion of complexity of crystals. More detailed treatment of the topic can be found in [47].

2.2.3.2 Orthogonal networks

Ideal orthogonal networks are networks with the angles between adjacent edges equal to 90 or 180° (adjacent edges are defined as having exactly one vertex in common). Each network that can be topologically (i.e., without breaking any edges) transformed into ideal orthogonal networks are orthogonal networks as well. The procedure of such a transformation is called an orthogonalization. Evidently, the maximal valency of a vertex in a three-dimensional orthogonal network is six and the full orthogonal network is the **pcu** net depicted in Fig. 2.4a. The **pcu** notation is taken from Delgado Friedrichs et al. [48] and denotes a *p*rimitive *cu*bic net. Any orthogonal framework can be obtained from the **pcu** net by deleting portion of its edges and/or nodes. Orthogonal networks may be zero-, one-, two, or three-dimensional.

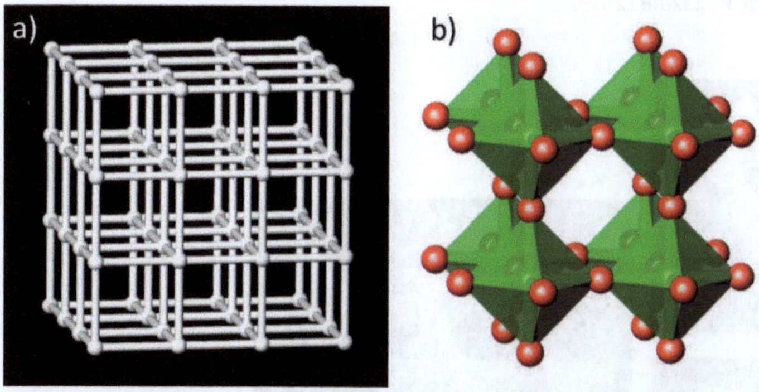

Fig. 2.4: The **pcu** net – a maximal three-dimensional orthogonal network (a) and perovskite-type framework of corner-sharing octahedra which topology can be described by the **pcu** net (b). Reproduced with the kind permission of the Mineralogical Society from [47], published in Mineralogical Magazine (2013).

The most straightforward realization of the **pcu** topology in inorganic compounds is the framework of corner-sharing octahedra present in the perovskite-type structures (Fig. 2.4b). Here each vertex of the **pcu** net corresponds to a single octahedron and each edge of the net corresponds to the linkage of two adjacent octahedra through a common O atom.

More than 50 % of the Earth's crust is composed from the minerals based upon two types of tetrahedral frameworks: feldspar (43.1 %) and quartz (11.9 %). Krivovichev et al. [46] demonstrated that the feldspar framework (Fig. 2.5) is orthogonal as well as tetrahedral frameworks in other minerals with feldspar stoichiometries. The tetrahe-

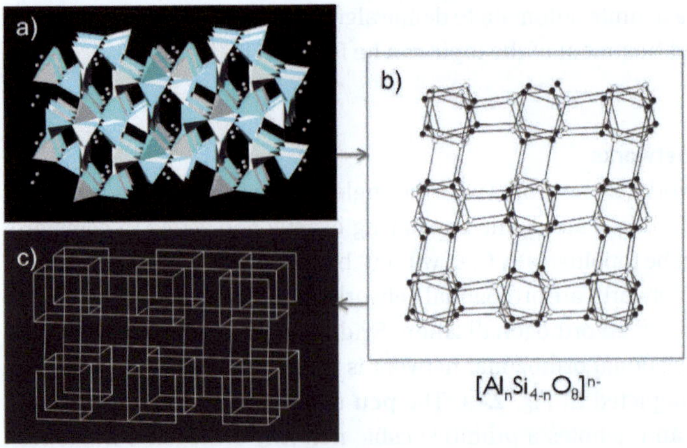

Fig. 2.5: The crystal structure of anorthite (a), nodal representation of its tetrahedral framework (black and white nodes correspond to AlO_4 and SiO_4 tetrahedra, respectively) (b), and its orthogonalized version (c). Reproduced with the kind permission of the Mineralogical Society from [47], published in Mineralogical Magazine (2013).

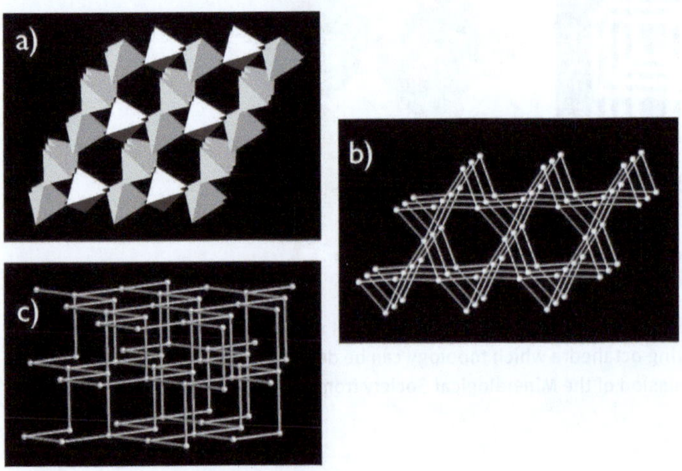

Fig. 2.6: Tetrahedral framework in the structure of quartz (a), its nodal representation (b) and orthogonalized version of quartz network (c). Reproduced with the kind permission of the Mineralogical Society from [47], published in Mineralogical Magazine (2013).

dral framework in the structure of quartz is shown in Fig. 2.6a. Its nodal representation (where each tetrahedron is symbolized by a vertex) is depicted in Fig. 2.6b as a three-dimensional graph with four-connected vertices. This graph can be orthogonalized and transformed into graph shown in Fig. 2.6c, which, in turn, can be obtained by ordered elimination of edges from the **pcu** network. Orthogonal networks can also be found in the structures of other rock-forming framework silicates such as cristobalite, tridymite, nepheline, kalsilite, leucite, etc.

2.2.3.3 Finite automata and formal languages: basic definitions and concepts

Orthogonal representation of complex networks as underlying topologies in crystal structures of inorganic compounds makes it possible to construct a simple deterministic model that generates periodic orthogonal networks. Since crystal structures are periodic, their generative models can be represented as finite automata. For the sake of simplicity, we use deterministic finite automata, though the elements of probability may be introduced to model disordered or partially ordered structures.

A deterministic finite automaton (DFA), A, is defined [49] as

$$A = \{Q, \Sigma, \delta, q_1, \mathbf{F}\},$$

where $Q = \{\mathbf{1}, \mathbf{2}, \mathbf{3}, \mathbf{4}, \ldots\}$ is the finite set of states, Σ is the finite set (*alphabet*) of transitional symbols (e.g., **a**, **b**, **c**, etc.), δ is a transition function, q_1 is an initial state, **F** is a set of accepting states from Q. The transition-function arguments are relevant to the current state and the transitional symbol; the result is a new state. For instance, the expression $\delta(q_1, \mathbf{a}) = q_2$ implies that, under the current state q_1 and the symbol **a**, the new state of the DFA will be q_2.

The structure of a DFA can be described using a *state diagram* and/or *transition matrix*. Fig. 2.7 shows the structure of the DFA consisting of four states (**1**, **2**, **3**, **4**) and three transitional symbols (**a**, **b**, **c**). The state diagram in the left part of the Fig. 2.7 shows how states of the DFA are changing depending upon the input transitional symbol. The same information is described by the transition matrix given in tabular form in the right part of the Fig. 2.7. The state **1** is chosen as an initial state. Under the input symbol **a**, the DFA is changing its state from **1** to **2**. Actually, the DFA in the state **1** accepts all three possible transitional symbols, **a**, **b**, and **c**. In contrast, the DFA in the state **3** accepts only the **b** and **c** symbols and does not accept the **a** symbol. This is because, for the state **3** and the symbol **a**, there is no new state defined by the state diagram or the transition matrix. The sequence of transitional symbols is defined as a *string* or a *word*. For instance, the DFA defined in Fig. 2.7 accepts strings **abcbc** and **bcccbc** and does not accept (rejects) strings **aba** or **bbc**. If **W** is the set of all words that the FDA A accepts, we say that **W** is the *formal language* of A and write $L(A) = \mathbf{W}$. We also say that A *recognizes* **W**. On the other hand, A generates **W** through the application of all possible pathways of changing states. The production rules (*formal grammar*) for the words from $L(A)$ are defined by either state diagram or transition matrix.

A simple example of a DFA is schematically shown in Fig. 2.8. It is an elevator in a house with three floors (Fig. 2.8a). It can be in one of three states (= floors): **1**, **2**, and **3**. The input system is inside the elevator and includes three buttons, which correspond to the numbers of the floors. Transition function is very simple: the elevator changes

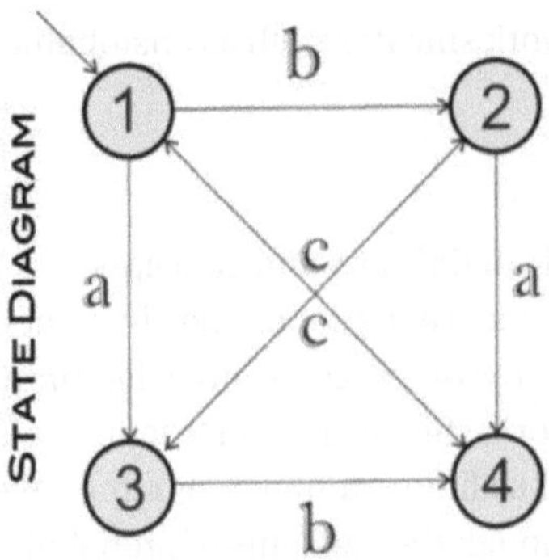

$Q = \{\mathbf{1}, \mathbf{2}, \mathbf{3}, \mathbf{4}\}$

$\Sigma = \{a, b, c\}$

TRANSITION MATRIX

$\delta =$

	a	b	c
1	**3**	**2**	**4**
2	**4**	-	**3**
3	-	**4**	**2**
4	-	-	**1**

$q_1 = \mathbf{1}$

Fig. 2.7: To the definition of deterministic finite automata. See text for details. Reproduced with the kind permission of the Mineralogical Society from [47], published in Mineralogical Magazine (2013).

its state according to the number of button pressed from inside the elevator. If the elevator is at the third floor (state **3**) and the button 1 is pressed on, the elevator moves to the floor 1 (state **1**). If the elevator is in full operation, any sequence of input numbers is accepted. The state diagram of the DFA that describes such an elevator is shown in Fig. 2.8b. However, let us suppose that the elevator is broken and it is not possible to get from the floor 1 to the floor 3 (and back) directly, but only with the stop at floor 2. The state diagram of such a 'broken' elevator DFA is shown in Fig. 2.8c. It is obvious that this DFA would not accept strings like 1123231223 or 223123, since they contain substrings 13 and 31, which cannot be recognized by the DFA. The DFA that recognizes

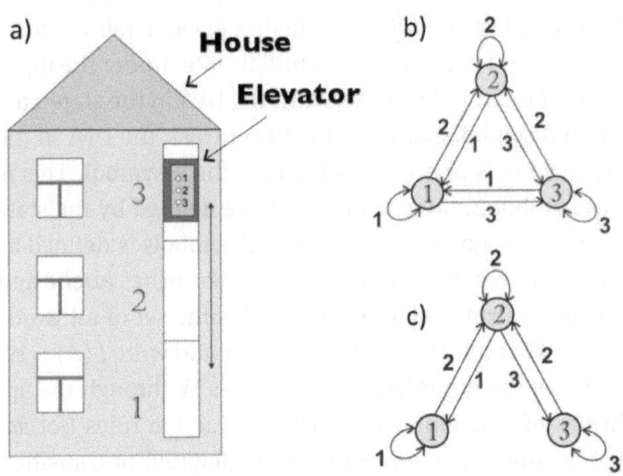

Fig. 2.8: Elevator as an example of a DFA (a): state diagrams for fully working (b) and broken (c) elevator systems. Reproduced with the kind permission of the Mineralogical Society from [47], published in Mineralogical Magazine (2013).

all possible sequences from the alphabet Σ is called an *universal automaton* A_U for the alphabet Σ.

2.2.3.4 Structural automata: basic constructions

For the generative description of crystal structures, Shevchenko et al. [45] and Krivovichev et al. [46] suggested to use *structural automata* (SA), which is the construction of DFA suggested by Morey et al. [50] for educational purposes in computer science. Fig. 2.9 shows correspondence between the elements of DFA and SA. A state corresponds to a vertex of a network with a certain configuration of adjacent edges. Symbols of the set Σ are vectors (directed edges). The transition function identifies a transition from state q_1 to state q_2 via a vector **v** that belongs to the set Σ: $\delta(q_1, \mathbf{v}) = k_2$. For each vector (symbol) **v**, there exists a vector $\underline{\mathbf{v}}$ such as $\delta'(q_2, \underline{\mathbf{v}}) = q_1$ (both **v** and $\underline{\mathbf{v}}$ belong to the set Σ). This means that structural automata are *bidirectional*. Any state from Q may be initial (q_1) and any state from Q is accepting.

Finite Automata	Structural Automata
State	Configuration of Vertex in the Network
Symbols	Vectors
Transition / Production Rules	Transition between Vertices through Vectors
Word	Vertex in the Network
Language	Network

Fig. 2.9: Correspondence between the elements of finite and structural automata. See text for details. Reproduced with the kind permission of the Mineralogical Society from [47], published in Mineralogical Magazine (2013).

To generate an orthogonal network, one needs six vectors (symbols) defined in Fig. 2.10a: $\Sigma = \{\mathbf{a}, \underline{\mathbf{a}}, \mathbf{b}, \underline{\mathbf{b}}, \mathbf{c}, \underline{\mathbf{c}}\}$. The vectors constitute an orthogonal basis in three-dimensional Euclidean space plus opposite vectors. The lengths of the vectors are equal to 1. Fig. 2.10b shows the state diagram of the structural automaton that generates the **pcu** network. The state diagram contains one state only: this is the consequence of the fact that all vertices in the **pcu** net are equivalent. All possible sequences of symbols from the alphabet Σ are accepted; the FDA that generates the **pcu** net, $A^{[\mathbf{pcu}]}$, is the universal automaton for the alphabet Σ.

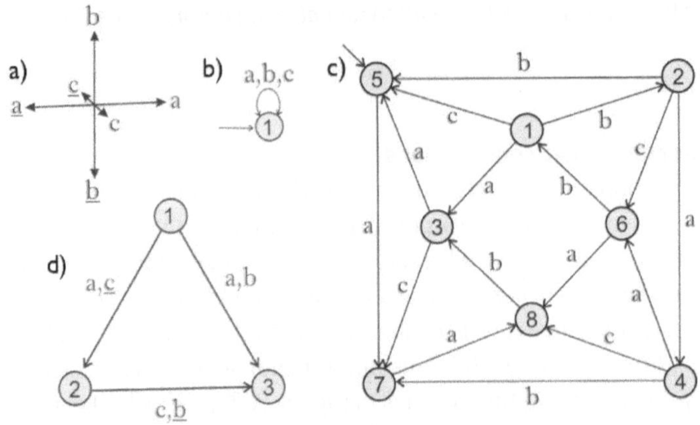

Fig. 2.10: Orthogonal vector basis (a) and state diagrams for structural automata that generate the **pcu** (b), quartz (c), and feldspar (d) nets. Reproduced with the kind permission of the Mineralogical Society from [47], published in Mineralogical Magazine (2013).

State diagrams for the structural automata that generate quartz and feldspar orthogonal networks (Figs. 2.5c and 2.6c, respectively) are shown in Figs. 2.10c and d, respectively. The quartz SA contains three states and is simpler than the feldspar SA that contains eight states.

The construction of the SA for a particular structure can be illustrated by the following example. Fig. 2.11a shows the two-dimensional tetrahedral layer from the structure of RUB-15, $[N(CH_3)_4]_8[Si_{24}O_{52}(OH)_4]\cdot 20H_2O$, a precursor phase in the formation of the sodalite-like microporous material [51]. Nodal representation of the layer is shown in Fig. 2.11b, whereas Fig. 2.11c provides its orthogonalized version. The network in Fig. 2.11c contains exactly three different vertex configurations (Fig. 2.11d). Therefore, the network-generating SA should contain three states that can be identified with three vertex configurations. Defining the **a** and **b** vectors as shown in Fig. 2.12c (right low corner), it is easy to construct the state diagram and the transition matrix for the corresponding SA (Fig. 2.12a,b). If the vertex λ is chosen as an initial vertex and the state **2** is assigned to it, the whole network can be produced by application of the respective production rules. New vertices of the net appear as the results of transition from the currently active vertices through the accepted vectors that play the role of transitional symbols in the SA. The generation of new vertices during the growth of the network is analogous to the generation of new words during the work of the respective FDA. Thus, any orthogonal framework can be put into correspondence with a certain and well-defined formal language. The production rules for the words of the language are dictated by the abstract structure of the corresponding FDA that may be expressed either as a state diagram (i.e. as a topological image) or as a transition matrix (i.e., as an algebraic structure). As all orthogonal networks are subnetworks of

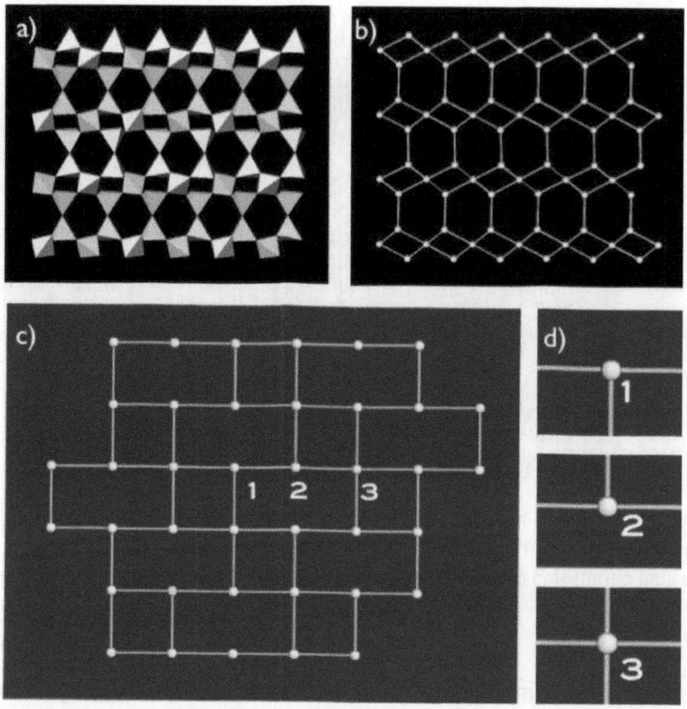

Fig. 2.11: Tetrahedral layer in the structure of RUB-15 (a), its nodal (b) and orthogonal (c) representations. The orthogonal network contains three different vertex configurations (d). Reproduced with the kind permission of the Mineralogical Society from [47], published in Mineralogical Magazine (2013).

the **pcu** net, all their languages are sublanguages of the universal language $L(A_U)$ for the alphabet $\Sigma = \{\mathbf{a}, \underline{\mathbf{a}}, \mathbf{b}, \underline{\mathbf{b}}, \mathbf{c}, \underline{\mathbf{c}}\}$.

2.2.3.5 Algorithmic complexity of structural automata

The dynamic model of growth of orthogonal networks by attachment of new vertices developed above allows formulating some criteria for the *algorithmic complexity* of orthogonal networks. Structural automata provide algorithms for the structure generation, and the complexity of these algorithms can be deduced from the analysis of the size and topology of state diagrams (it is very probable that other measures can be proposed through the analysis of complexity of transition matrices). As the first and more natural measures of algorithmic complexity one can use the number of states of the SA.

The interplay between dynamic and static complexity of crystal structures is an interesting problem that requires additional research. The crystal structures of cristobalite and quartz are based upon the **dia** and **qtz** (quartz) networks, respectively.

a)

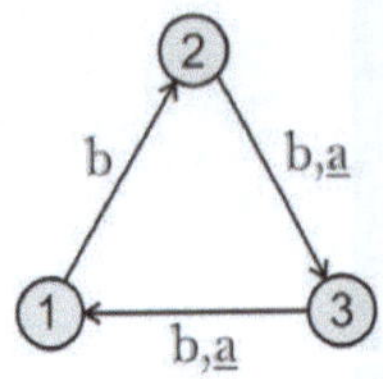

b) Transition Matrix

	a	a	b	b
1	3	-	2	3
2	-	3	3	1
3	2	1	1	2

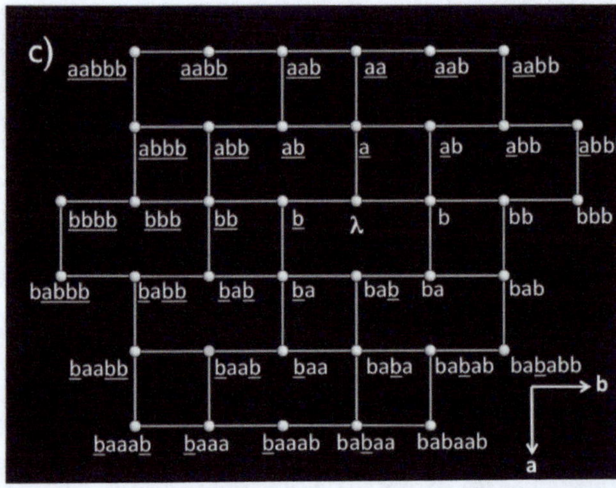

Fig. 2.12: To the description of finite automaton generating the orthogonal RUB-15 network: state diagram (a), transition matrix (b), and portion of the network with words assigned to the vertices (c). Reproduced with the kind permission of the Mineralogical Society from [47], published in Mineralogical Magazine (2013).

Their algorithmic complexities defined as the number of states in the SA are equal to two and three, respectively. The static complexities of the crystal structures of cristobalite and quartz defined as structural information amounts [15] are equal to 5.510 and 8.265 bits/cell (per unit cell), respectively. Algorithmic complexity of the tridymite tetrahedral framework is equal to four, whereas its static complexity is 17.510 bits/cell Therefore, both dynamic (algorithmic) and static complexities for the three SiO_2 polymorphs can be described by the same relation: the tridymite topology is more complex than the quartz topology, which is more complex than the cristobalite topology.

Krivovichev et al. [46] applied the model of structural automata to investigate complexity of crystal structures of the $CaAl_2Si_2O_8$ polymorphs, including those that appear during metastable crystallization. Orthogonal frameworks of both metastable polymorphs, svyatoslavite and dmisteinbergite, can be generated by SA with four states, whereas state diagram for the feldspar framework contains eight states. These relations are corroborated by the static complexities of anorthite, svyatoslavite and dmisteinbergite that are equal to 76.211, 62.039 and 35.020 bits/cell, respectively.

2.3 Information-based structural complexity of minerals: Statistics and basic trends

A total of 19531 structure reports on minerals was extracted from the Inorganic Crystal Structure Database (ICSD, version 2011-2) and analyzed using the crystal-chemical program package TOPOS [52]. From these, 3949 entries were selected according to the following criteria: (a) for the same mineral, the most recent and most complete structure refinement was chosen for final calculations (e.g. a structure report with determined H atoms was preferred over the report with no H atoms); (b) if the same mineral was refined in different space groups, the entries for each space group were used.

Fig. 2.13 shows the total structural information content of minerals ($I_{G,\text{total}}$) versus the information content per atom (I_G) and the number of atoms in the reduced unit cell (v). The graphs show that neither the symmetry-sensitive I_G, nor the size-sensitive v parameters can be used as single complexity measures, whereas the combined $I_{G,\text{total}}$ value is sensitive to both size and symmetry.

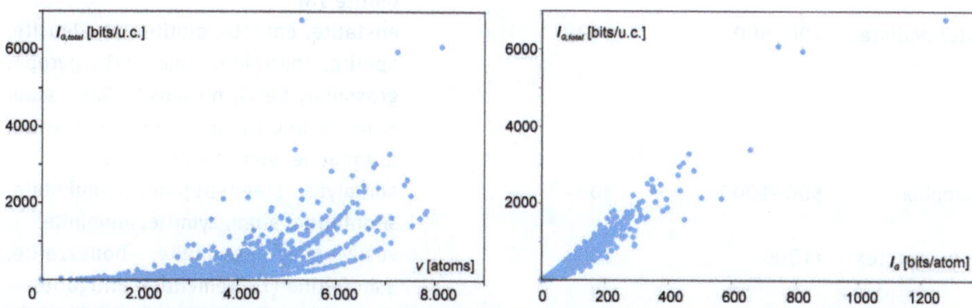

Fig. 2.13: The total structural information content of minerals ($I_{G,total}$) plotted against number of atoms in the reduced unit cell (v; above) and the information content per atom (I_G; below). Reproduced with the kind permission of the Mineralogical Society from [15], published in Mineralogical Magazine (2013).

Statistical analysis of the data on structural complexity of all known minerals allowed us to calculate the average information content for mineral structures as 228(6) bits per structure and 3.23(2) bits per atom. A classification scheme for minerals according to their structural information content or complexity is presented in Tab. 2.1. Minerals can be subdivided into 5 categories: [1] very simple, [2] simple, [3] intermediate, [4] complex, and [5] very complex. Most of the rock-forming minerals fall into the first three categories. However, several exceptions are known. For instance, vesuvianite, a common mineral in skarns, is regarded as very complex. The structural complexity of vesuvianite is a consequence of its modularity: its structure is built from one-dimensional modules from the structure of grossular [53]. Channels between the modules are occupied by a range of cations and anions, which results in a rather complex crystal-chemical behavior [54–56].

According to their crystal chemistry, most complex mineral structures can be subdivided into zeolites and complex feldspathoids, misfit layered silicates, minerals with nanoscale clusters and tubules, complex sulfates (apjohnite), and complex modular structures. A more detailed overview of selected examples is given in [15].

Tab. 2.1: Classification of minerals according to their complexity [15].

Category	Total information content [bits/cell]	Approximate number of mineral species	Examples
very simple	0–20	600	diamond, copper, halite, galena, uraninite, fluorite, quartz, corundum, ringwoodite, calcite, dolomite, zircon, goethite, lepidocrocite
simple	20–100	1100	alunite, jarosite, nepheline, kieserite, szomolnokite, kaolinite, olivine-group minerals, diopside, orthoclase, albite, biotite $1M$
intermediate	100–500	1800	enstatite, epidote, biotite $2M_1$, leucite, apatite, natrolite, talc $2M$, pyrope, grossular, beryl, muscovite $2M_1$, staurolite, actinolite, holmquistite, coesite, tourmaline, analcime, boracite
complex	500–1000	300	eudialyte, steenstrupine, coquimbite, sapphirine, alum, cymrite, aluminite
very complex	>1000	100	vesuvianite, paulingite, bouazzerite, ashcroftine-(Y), bementite, antigorite

2.4 Structural complexity and scalar hierarchy

The term 'structural hierarchy' was introduced into structural mineralogy by P. B. Moore [57], who defined it as 'a general scheme which, through the appropriate algorithms, ties together a collection of arrangements'. However, implicitly this concept has been in use among mineralogists and crystallographers since the landmark papers by Machatschki [58] and Bragg [59] on the crystal chemistry of silicates. Very recently, Hawthorne [60] reviewed the development of a structure hierarchy hypothesis since 1920s and defined it as 'a classification of atomic arrangements arranged (ranked) according to their principal constituent cation-polyhedra and the connectivity of those polyhedra'. The principal cation-polyhedra implied by this definition are coordination polyhedra with higher bond-valences, following the original idea proposed in [25].

In fact, most of the known structural classifications of minerals imply the construction of structural hierarchy schemes based upon specific criteria such as dimensionality that specify the respective hierarchical levels (see, e.g., systematics of silicates by Liebau [61]).

In general, the modern science (or the modern philosophy of science) distinguishes between two different types of hierarchy [62]. The first type is a subsumption or *specification* hierarchy such as the Linnaean hierarchy of biological species. This form of hierarchy has its levels as nested subclasses and, in mineralogy, has a usual form of structural-chemical systematics of minerals such as the one developed by Strunz and Nickel [63]. An example of a specification hierarchical analysis applied to minerals is

{minerals {silicates {framework silicates {zeolites {paulingite }}}}}.

Here the highest hierarchy level is represented by minerals in general, i.e. crystalline compounds of natural origin, and the lowest level is a particular mineral, paulingite. The term 'structural hierarchy' used by Moore [57] and the concept 'structure hierarchy' reviewed recently by Hawthorne [60] deals with the kind of specification hierarchy that is based upon particular features of structural arrangements (dimensionality, connectivity and topology of structural units, for example). Graphically, this type of hierarchical relations is depicted as cladograms and nested trees.

Another kind of hierarchy is a compositional hierarchy, also known as a *scale* or inclusion hierarchy. It considers a specific system as consisting of subsystems (components) nested within the system (whole). The example is an electron inside an atom inside a coordination polyhedron inside a structural unit inside a unit cell inside a crystal:

{crystal {unit cell {structural unit {coordination polyhedron {atom {electron }}}}}}.

This kind of a hierarchy is implied in any structural classification of minerals and also has a high relevance for the related concept of modularity. The quantitative basis for the compositional hierarchical analysis was established by Hawthorne [25] by means of the bond-valence theory. The approach used by Hawthorne [25] is coherent with the idea of a nearly decomposable system developed by Simon [64] in his famous paper 'The Architecture of Complexity'. Such a system can be decomposed into subsystems according to the strengths of interactions of subsubsystems within them.

Simon [64] noticed that 'hierarchy is one of the structural schemes that the architect of complexity uses'. Fig. 2.14 shows a general scheme of a hierarchical organization of a crystal structure of a mineral. In general, five hierarchical levels can be identified (from down to top):

(1) atoms as primary basic units (PBUs);
(2) coordination polyhedra as basic building units (BBUs);
(3) fundamental building blocks (FBBs) or composite building units (CBUs);
(4) structural units and interstitial complexes;
(5) structure as a whole.

It is obvious that not all minerals are organized in such a way to possess all five possible levels of a scale hierarchy. Therefore, the number of hierarchical levels or a *hierarchical depth* (HD), can be considered as a measure of structural complexity. Another measure is a *hierarchical span* (HS) that reflects the size and diversity of organized entities on different hierarchy levels.

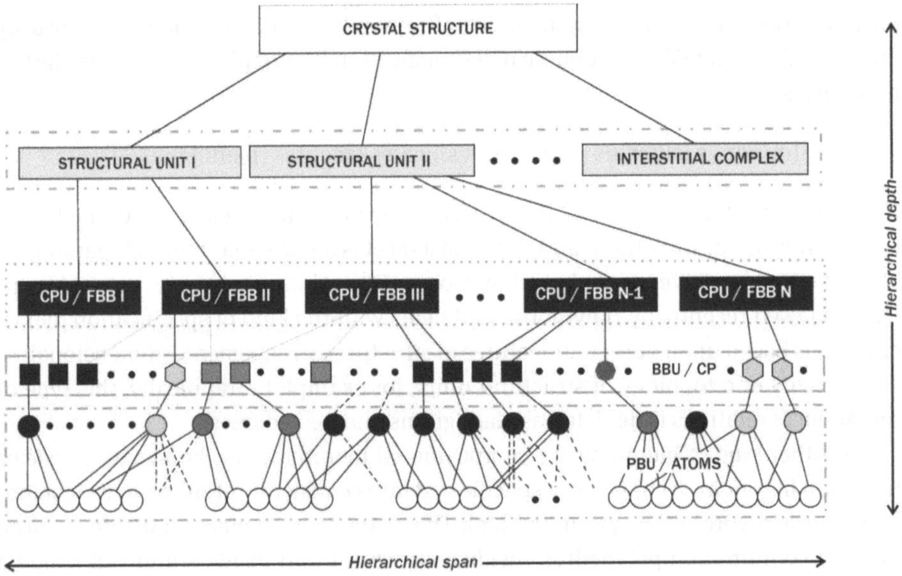

Fig. 2.14: General scheme of hierarchical organization of an inorganic crystal structure.

For instance, the crystal structure of native copper contains only two hierarchical levels (Fig. 2.15a) and its hierarchy is very shallow, which underlines its structural simplicity. The hierarchical depths of the crystal structures of quartz (Fig. 2.15b) and albite (Fig. 2.15c) are three and four, respectively, which agrees well with the intuitive feelings of their structural complexity. A high complexity of the crystal structure of charoite can be seen from its structural hierarchy scheme shown in Fig. 2.16. The hierarchical organization of charoite contains five hierarchical levels (HD = 5) and its hierarchical span is really spectacular. Even on the third level, it contains five different CBUs, three of which belong to three topologically different types of one-dimensional silicate anions.

However, the difficulty with quantifying complexity in terms of hierarchical depth and span lies in the sometimes subjective subdivision of hierarchy levels as it was shown by Ferraris et al. [65]. As it was pointed out by Salthe [62], ‘only some users of hierarchical forms would insist that particular levels actually exist as such in natural systems’, as hierarchy is just a method of ‘cutting (a seemingly continuous) Nature at its joints.’ However, analysis of a hierarchical organization of crystal structures of minerals agrees well with the quantitative analysis of their complexity in terms of their Shannon information amount [14–16].

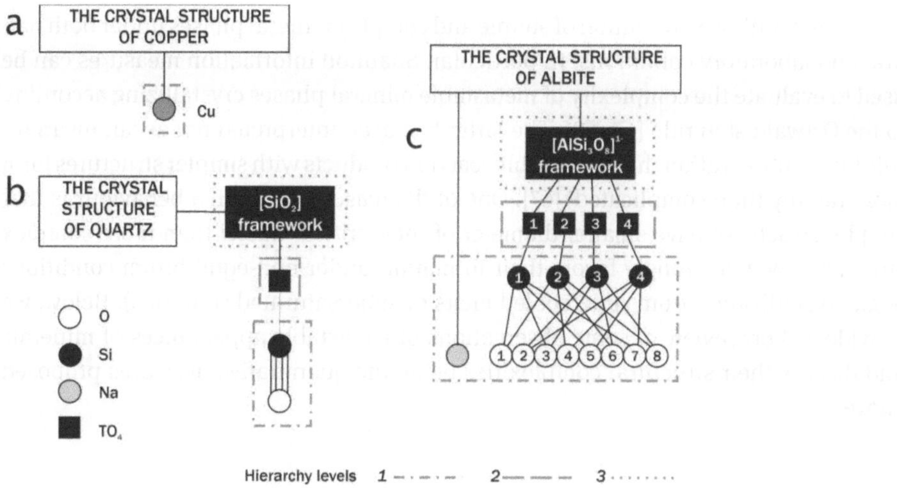

Fig. 2.15: Examples of compositional hierarchy schemes of the crystal structures of minerals: native copper (a), quartz (b) and albite (c).

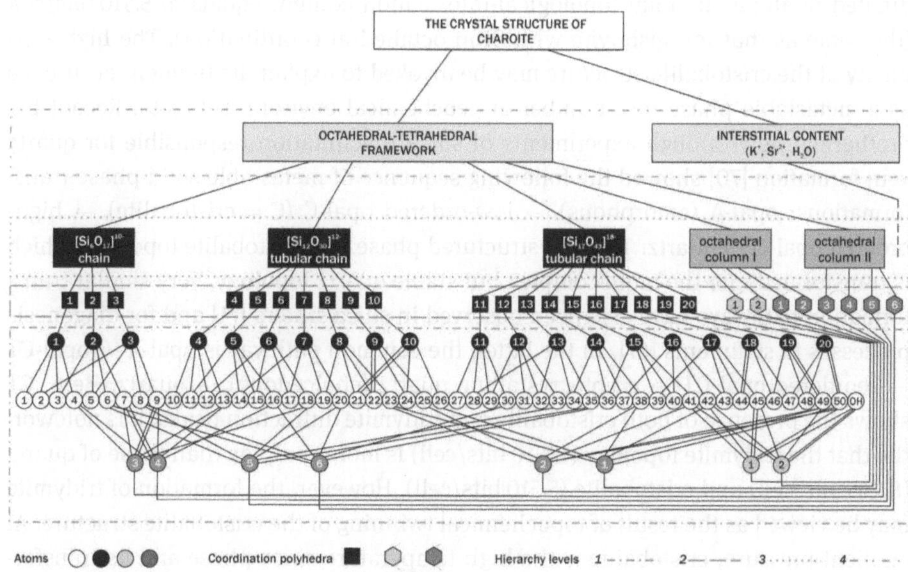

Fig. 2.16: Compositional hierarchy scheme for the crystal structure of charoite.

2.5 Structural complexity and the 'ease of crystallization'

2.5.1 General notes

The information-based complexity measures can be used to estimate the 'ease of crystallization' from the viewpoint of the simplexity principle proposed by Goldsmith [66] for understanding of formation of simple and complex mineral phases under both natural and laboratory conditions. In particular, Shannon information measures can be used to evaluate the complexity of metastable mineral phases crystallizing according to the Ostwald step rule [67, 68]. The latter has been interpreted not as 'an invariant rule but an observation that metastable reaction products with simpler structures form more rapidly than complicated' [67]. One of the reasons for such a behaviour is that simpler structures have smaller diameter of their critical nuclei than more complex structures, which strongly favors their formation under non-equilibrium conditions (e.g., crystallization from supercooled melts or supersaturated solutions). Below, we provide a short review of some observations of metastable appearances of minerals and discuss their structural complexities using the quantitative measures proposed above.

2.5.2 Silica polymorphs

Cristobalite is the topologically simplest silica polymorph with tetrahedrally coordinated Si atoms. Its total topological-information content equals to 5.510 bits/cell (the same as that for stishovite with Si in octahedral coordination). The high simplicity of the cristobalite structure may be invoked to explain its frequent occurence as a metastable phase in a number of geochemical environments [69]. Recent hydrothermal flow-through experiments of silica precipitation responsible for quartz vein formation [70] showed the following sequence of metastable solid-phase transformations: opal-A (amorphous) → low-ordered opal-C (C = cristobalite) → high-ordered opal-C → quartz. The first structured phase has cristobalite topology, which is transformed into the more complex but stable quartz topology. Very similar transformation pathways have also been observed in silica sinters [71] and for diagenesis processes in sediments [67]. In the latter, the common pathway is opal-A → opal-CT → reordered opal-CT → cryptocrystalline quartz (chalcedony) → quartz. Here, CT shows the presence of both cristobalite and tridymite diffraction peaks. It is noteworthy that the tridymite topology (17.510 bits/cell) is more complex than those of quartz (8.265 bits/cell) and cristobalite (5.510 bits/cell). However, the formation of tridymite may be viewed as the result of topochemical twinning of the cristobalite structure. At ambient pressure, cristobalite is the high-temperature silica phase and its transformation to quartz with decreasing temperature proceeds via the intermediate tridymite structure, which means that structural information increases along the pathway. It is very likely that the occurence of tridymite as a transitional phase may be explained by purely structural reasons, i.e. a steric convenience for the reconstruction of the cristobalite topology into that of quartz.

2.5.3 Feldspar compositions

Svyatoslavite [72] and dmisteinbergite [73] are two metastable polymorphs of $CaAl_2Si_2O_8$, which crystallize in the CaO-Al_2O_3-SiO_2 system prior to the crystallization of anorthite. They were first identified by Davis and Tuttle [74], and their stability, nucleation and growth kinetics were investigated in [75–77]. Abe et al. [75] reported that the pseudo-hexagonal (dmisteinbergite) and pseudo-orthorhombic (svyatoslavite) polymorphs nucleate prior to anorthite and grow in a supercooled $CaAl_2Si_2O_8$ melt until anorthite starts to crystallize. At that moment, these phases either dissolve or transform to anorthite. As the structure of dmisteinbergite is not precisely determined (there is information on its synthetic counterpart only [76, 77]), direct comparison of structural complexity of the metastable polymorphs and anorthite is not possible. However, the topology of the aluminosilicate frameworks is well-established for all polymorphs. The dmisteinbergite structure is based on double sheets of six-membered rings of AlO_4 and SiO_4 tetrahedra and has the lowest topological information content (35.020 bits/cell). Both svyatoslavite and anorthite are based on three-dimensional frameworks of tetrahedra, but the svyatoslavite topology is simpler (62.039 bits/cell) than that of anorthite (76.211 bits/cell). Therefore, both metastable $CaAl_2Si_2O_8$ polymorphs are simpler than the stable anorthite structure, which explains their spontaneous crystallization from a supercooled melt. This is also an obvious explanation for their crystallization from high-temperature (up to 1200 °C) gases in the coal dumps of the Chelyabinsk coal area [72, 73]. Nestola et al. [80] reported metastable formation of dmisteinbergite in a pseudotachylyte by rapid quenching of a friction melt generated by seismic slip. They suggest that 'given the small grain size, <...>dmisteinbergite may not be so rare, but rather that its abundance is simply underestimated... dmisteinbergite is probably more common than reported in rocks produced during large thermal impulses.' Very recently, Ma et al. [81] reported on metastable formation of dmisteinbergite in the Allende meteorite, where it could have crystallized from a silicate melt at high temperature (~1200–1400 °C) via rapid cooling.

Hwang et al. [82] reported on kumdykolite, an orthorhombic (most probably pseudo-orthorhombic) polymorph of albite, $NaAlSi_3O_8$, which has very similar unit-cell parameters to svyatoslavite [46]. Hwang et al. [82] pointed out that kumdykolite, discovered in the Kokchetav ultrahigh-pressure massif in Kazakhstan, is a metastable phase formed at high temperature by rapid cooling in the absence of H_2O. If kumdykolite has the svyatoslavite topology (which seems likely), its formation may be explained by the higher simplicity of its structure topology compared to that of albite. Neméth et al. [83] discovered kumdykolite in enstatite chondrite Sahara 97072, EH3, and proposed that the mineral formed above 1300 K and cooled rapidly enough to preserve its unique structure.

Kokchetavite, $KAlSi_3O_8$, is another mineral from Kokchetav massif [84], which most probably formed metastably in the stability field of K-feldspar from infiltrated melts during retrograde exhumation. Its structure is based on double aluminosilicate sheets with the dmisteinbergite topology, which is simpler than the stable feldspar framework-topology. It is of interest that kokchetavite occurs in association with cristobalite, which supports its metastable appearance as a phase that is simpler than its stable counterparts.

The dmisteinbergite double-sheet topology has also been observed in 'hexacelsian', $BaAl_2Si_2O_8$ (not a valid mineral species), cymrite, $BaAl_2Si_2O_8(H_2O)$, and 'K-cymrite', $KAl_2Si_2O_8(H_2O)$ [85]. Both 'hexacelsian' and cymrite crystallize metastably well below their stability field [86–88], within the stability field of celsian. One of the reasons for their preferential crystallization may be derived from the lower complexity of their structure topologies.

2.5.4 Serpentine-group minerals

Antigorite is the stable phase in olivine-serpentine reactions [89, 90], but its nucleation rate is sluggish [91], which is usually explained by the high complexity of its structure. Dungan [92] pointed out that 'the role of crystal structure is manifested in the relative ease of nucleation of the various phases' and that 'antigorite nucleation is hampered by its complex structure'. Indeed, the structural information content for different antigorite polysomes varies from 1277 to 2188 bits/cell [93–96], which corresponds to very complex structures according to the classification proposed above (Tab.2.1). In contrast, crystal structures of chrysotile and lizardite are much simpler (43–86 and 35–112 bits/cell for different polytypes, respectively, and that favors their dominant (and, in the case of chrysotile, metastable [90]) formation in serpentinites. The structures of all serpentine-group minerals are based on sheets of silicate tetrahedra and continuous sheets of $Mg(O,OH)_6$ octahedra. The misfit between the tetrahedral and octahedral substructures results in strain, which is relaxed by rotation of tetrahedra in lizardite, formation of rolls and tubes in chrysotile, or inversion of tetrahedra in antigorite. Different numbers of inverted tetrahedra result in different polysomes, which are characterized by the number m of tetrahedra along the a repeat. According to Mellini et al. [97], a typical range of m values is 13–24. The structural complexity of antigorite polysomes is a function of m. For instance, the structural information contents for the $m = 16$ and $m = 17$ polysomes [95, 96] are 1761.323 and 2187.799 bits/cell, respectively. Wunder et al. [91] investigated pressure and temperature dependence of polysomatism and H_2O content in antigorites and found that m decreases with increasing temperature and decreasing pressure. This is in agreement with the general trends of structure complexity increasing with decreasing temperature and moderately increasing pressure. Under ultrahigh-pressures, the situation may be drastically different (see above).

2.5.5 Micas and clay minerals

It has been shown by many observations and experimental studies that the first phase that forms during initial stages of crystallization of muscovite and illite is either the $1M$ or $1M_d$ polytype ($1M_d$ has a strongly disordered structure). Velde [98] proposed that the stable polytype of dioctahedral white mica is $2M_1$, whereas the $1M$ and $1M_d$ polytypes are metastable. Mukhamet-Galeyev et al. [99] synthesized muscovite from stoichiometric-composition gels and showed that the $1M$ and $2M_1$ polytypes form at the same time. However, the crystallization rate of the $1M$ polytype is much higher (it crystallizes *more easily*) and complete crystallization of the gel occurs almost completely in the $1M$ form. After this, the $1M$ form starts to convert to the $2M_1$ form, but the conversion rate is very slow. According to Meunier and Velde [100], this explains why the 1*M* polytype occurs in natural diagenetic sequences where it is a metastable form of illite. Baxter Grubb et al. [101] explained the formation of the metastable $1M$ and $1M_d$ polytypes from the viewpoint of the Ostwald step rule. The relative ease of crystallization of the 1*M* polytype is in agreement with the Goldsmith's simplexity principle: the structure of the $1M$ polytype is essentially simpler than that of the $2M_1$ polytype. For muscovite, their information contents are equal to 60.239 and 136.877 bits/cell, respectively.

2.5.6 Metastability and structural complexity

The observations given above indicate that structural complexity may have a profound impact upon mineral reactions when kinetic factors are of importance and the system does not have enough energy to overcome the energy barrier that separates metastable simple and stable complex structures. Under non-equilibrium conditions, simpler structures may form metastably in the stability fields of more complex structures. However, metastable structures are not always the simplest ones. For instance, zeolites and other microporous materials, many of which are complex or very complex, are generally metastable phases in the corresponding chemical systems [102].

By rapid cooling the Pb-O-Br-Cl melt with the mendipite-like composition, $Pb_3O_2(Cl,Br)_2$, Krivovichev et al. [103] obtained the compound $Pb_{31}O_{22}X_{18}$ (X = Cl, Br), which has one of the most topologically complex tetrahedral layers known so far. Its structure is much more complex (873.264 bits/cell) than its stable mendipite-like counterpart (70.606 bits/cell). In many cases, the high complexity of metastable structures can be explained by their inheritance of the structures of starting reagents of chemical reactions. In particular, the crystal structure of $Pb_{31}O_{22}X_{18}$ (X = Cl, Br) is based upon complex $[O_{22}Pb_{31}]^{18+}$ layers that are direct derivatives of the crystal structure of PbO, which, along with $PbCl_2$ and $PbBr_2$, was the starting reagent of the reaction. Therefore, the complexity of metastable state is due to the transitional character of its structure that occurs in between more structurally simple initial and final stages.

The interesting case was observed for the crystal structure of β-Bi(SeO_3)Cl, which occurs as a transitional phase between the α and γ polymorphs and exists in a very narrow temperature range [104]. The complexity of the α and γ phases is the same and equals to 62.039 bits/cell (that brings them to the category of simple structures: see Tab. 2.1), whereas that of the transitional β phase is 4705.880 bits/cell (very complex structure). The analysis of the crystal structures of the Bi(SeO_3)Cl polymorphs in terms of Bi cation arrays [104] indicated that the cation array of the β phase is intermediate between those of the α and γ phases. This shows that highly complex structures may appear as transitional states along reconstruction pathways between simple structures.

2.6 Microevolution of structural complexity in natural processes

The information-based complexity measures can be used to investigate evolution of paragenetic associations of minerals in a particular geochemical environment. Krivovichev [15] investigated the evolution of complexity of minerals during hydrothermal reworking of primary phosphates following Moore [105] and Hawthorne [106] to demonstrate that complexity is increasing with the decreasing temperature and increasing chemical complexity of the system. Below we provide two other examples of similar analysis that may lead to some interesting and important conclusions.

2.6.1 Information and chemical reactions: oxidation of pyrite and the cascade of iron sulfate hydrates

The discovery of iron sulfates on the Mars surface (see [107] and references therein) renewed the interest in formation and transformation of these minerals under conditions of changing temperature and humidity [108–110]. On the Earth, iron sulfates are among the most common secondary phases that form in oxidation zones of sulfide mineral-deposits and as corrosion products of steel. Jambor et al. [111] provided a detailed review of paragenetic sequences of Fe sulfate minerals resulting from pyrite oxidation. The scheme of transformations of different phases based upon both field studies and laboratory experiments is summarized in Fig. 2.17. Here each phase is characterized by its topological diagram that shows the topology of linkage between [$Fe\phi_6$] octahedra (ϕ = O^{2-}, OH^-, H_2O) and SO_4 tetrahedra in terms of a black-and-white graph as elaborated for sulfates by Hawthorne et al. [112]. As it can be seen, the transformation pathways between the phases are quite complex with a possibility of reverse transformations depending upon humidity and temperature conditions. However, a very general sequence of transformations starting from pyrite and ending at goethite can be established as follows:

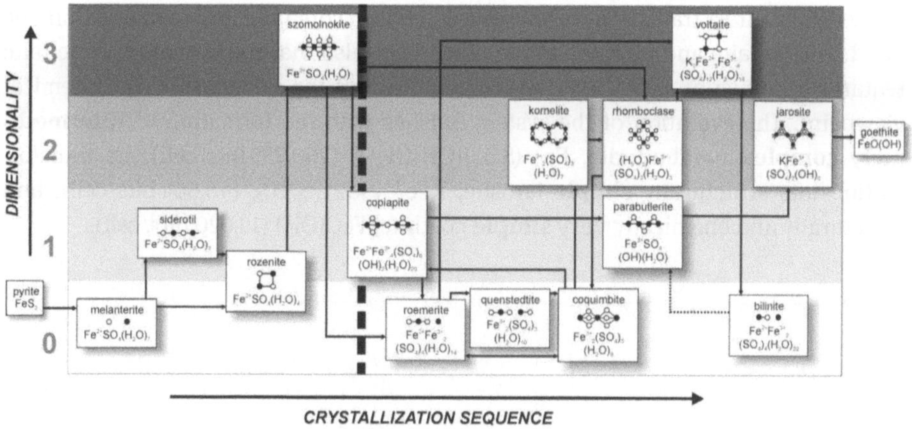

Fig. 2.17: General scheme of phase transformations occurring in oxidation zones of Fe silfide mineral deposits.

pyrite → melanterite → siderotil → rozenite → szomolnokite → copiapite → roemerite → quenstedtite → coquimbite → kornelite → rhomboclase → voltaite → bilinite → parabutlerite → jarosite → goethite

Fig. 2.18 shows the evolution of structural information per unit-cell along this sequence of phases. As it may be expected, the evolution is non-linear and demonstrates an oscillating behavior. However, some interesting features of this cascade of transformations can be revealed by more detailed analysis. The initial destruction of the **very simple** structure of pyrite (11.020 bits/cell) is accompanied by the oxidation reaction $S^{2-} \rightarrow S^{6+}$, the formation of the sulfate oxyanion, SO_4^{2-}, and the release of energy. The first sulfate phase that forms in the system is melanterite, $Fe^{2+}(SO_4)(H_2O)_7$, which has a **complex** structure (517.528 bits/cell). It transforms into siderotil, $Fe^{2+}(SO_4)(H_2O)_7$ (184.477 bits/cell), rozenite, $Fe^{2+}(SO_4)(H_2O)_4$ (300.235 bits/cell), and szomolnokite, $Fe^{2+}(SO_4)(H_2O)$ (45.059 bits/cell), which corresponds to the decrease of structures from **complex** to **intermediately complex** and back to **simple.** The oxidation of Fe^{2+} to Fe^{3+} provides further input of energy into the system, which allows to build such mixed-valent Fe^{2+}-Fe^{3+} sulfates as copiapite, $Fe^{2+}Fe_4^{3+}(SO_4)_6(OH)_2(H_2O)_{20}$ (558.306 bits/cell), and roemerite, $Fe^{2+}Fe_2^{3+}(SO_4)_4(H_2O)_{14}$ (327.454 bits/cell), which are classified as **complex** and **intermediately complex**, respectively. These phases may transform into purely Fe^{3+} sulfates such as quenstedtite, $Fe_2^{3+}(SO_4)_3(H_2O)_{10}$ (269.212 bits/cell), coquimbite, $Fe_2^{3+}(SO_4)_3(H_2O)_9$ (635.384 bits/cell), kornelite, $Fe_2^{3+}(SO_4)_3(H_2O)_7$ (691.895 bits/cell), and rhomboclase, $(H_5O_2)Fe^{3+}(SO_4)_2(H_2O)_3$ (336.168 bits/cell), that can be described as fluctuations of structural state of minerals between **intermediate** and **complex** levels. The reverse $Fe^{2+} \leftrightarrow Fe^{3+}$ reactions result in formation of the **very complex** structures of voltaite, $K_2Fe_5^{2+}Fe_4^{3+}(SO_4)_{12}(H_2O)_{18}$ (1566.017 bits/cell), and bilinite, $Fe^{2+}Fe_2^{3+}(SO_4)_4(H_2O)_{22}$ (2305.361 bits/cell). Fig. 2.17

shows that a direct transition was observed between copiapite and roemerite, on one hand, and voltaite and bilinite, on the other. Note also that the formation of voltaite requires introduction of K^+ ions, which complexifies the system from the chemical viewpoint. The evolution of the system finishes with the formation of **intermediately complex** parabutlerite, $Fe^{3+}(SO_4)(OH)(H_2O)$ (260.235 bits/cell), its transformation into structurally **simple** jarosite, $KFe^{3+}(SO_4)_2(OH)_6$ (46.664 bits/cell), and structurally and chemically **very simple** goethite, FeO(OH) (19.020 bits/cell).

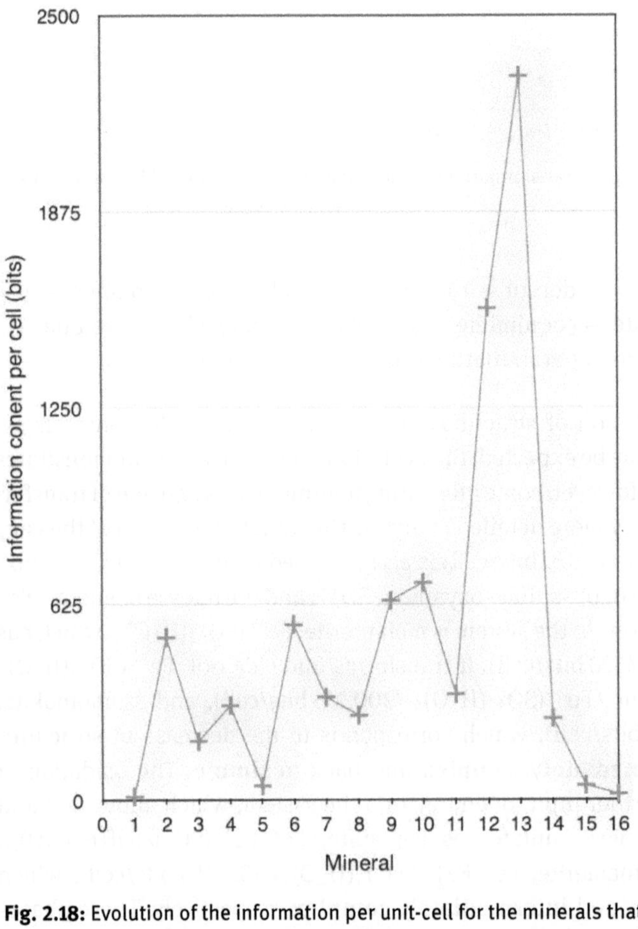

Fig. 2.18: Evolution of the information per unit-cell for the minerals that form during oxidation of pyrite. Legend for the X axis: 1 – pyrite, 2 – melanterite, 3 – siderotil, 4 – rozenite, 5 – szomolnokite, 6 – copiapite, 7 – roemerite, 8 – quenstedtite, 9 – coquimbite, 10 – kornelite, 11 – rhomboclase, 12 – voltaite, 13 – bilinite, 14 – parabutlerite, 15 – jarosite, 16 – goethite.

In terms of structural complexity, the behavior of the iron sulfate system in oxidation zones of mineral deposits can be interpreted as an energy-for-information exchange.

The input of energy into the system due to the oxidation reactions results in the rise of complexity, whereas, in the absence of such an input keeps the system at the level of complexity achieved at the previous stage. Very recently, Christy et al. [113] analyzed structural and chemical evolution of Te mineral parageneses at Bird Nest drift, Otto Mountain, California, USA, and described very detailed paragenetic sequences of Te minerals occurring at that locality. Analysis of evolution of structural information in the generalized paragenetic scheme did not reveal any particular trends with the variations in complexity from 21 to 416 bits per cell (i.e. within ~400 bits/cell). It can be compared to the variations observed in the Fe sulfate system with the differences in information of separate phases reaching 2000 bits/cell). The most probable reason for the rather shallow behavior of information observed in the Bird Nest drift mineral parageneses is the absence of any considerable changes in either temperature or energy of the system. In addition, the chemistry of the parageneses is not constant and much more diverse than that of the Fe sulfate system, with the number of different chemical elements in the range from 3 to 7. In general, it may be concluded that, in the information-based analysis of paragenetic evolution of mineral systems, it is better to focus on those associations or transformation sequences, which are more or less isolated from the chemical point of view.

2.6.2 Information and temperature: Natural zeolites

Krivovichev [114] analyzed structural and topological complexities of zeolites and zeolite frameworks using the information-theory methods as described above. The obtained values of topological information for different tetrahedral frameworks may be used to investigate evolution of complexity in zeolite parageneses. As it has already been mentioned, in such an analysis, it is better to concentrate on really or hypothetically chemically isolated systems. In fact, there are only very few studies on such systems available in the literature. Bargar and Keith [115] provided an overview of the research that has been conducted using drill holes in active hydrothermal areas at Yellowstone National Park, USA. It has been observed that zeolites are among the last minerals to form and each zeolite has its own temperature limits of stability. For instance, erionite was found only at temperatures below ~100 °C, whereas analcime was found at temperatures > 80 °C, etc.

Chipera and Apps [116] provided thermodynamic modeling of the zeolite parageneses in silicic-rock hydrothermal environments that contained a representative suite of zeolites (analcime, chabazite, clinoptilolite, erionite, heulandite, laumontite, mordenite, phillipsite, stilbite, and wairakite). Fig. 2.19 shows the $\log[(aK^+)^2/aCa^{2+}]$ vs. $\log[(aNa^+)^2/aCa^{2+}]$ diagrams constructed by Chipera and Apps [116] for the temperatures of 25, 100, 150, and 200 °C. First, it is of interest that, in terms of diversity, the system at 100 °C appears the richest (10 phases), whereas the system at 200 °C is the poorest (4 phases). In order to analyze **structural** complexity of the parageneses at

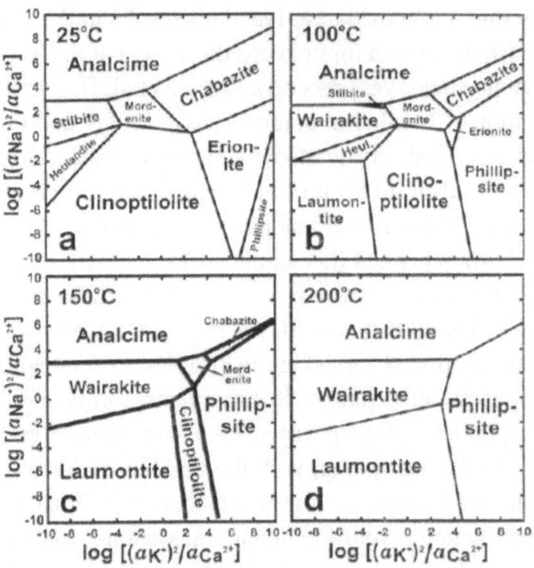

Fig. 2.19: Log[$(a\text{K}^+)^2/a\text{Ca}^{2+}$] *vs.* log[$(a\text{Na}^+)^2/a\text{Ca}^{2+}$] diagrams for zeolites occurring in hydrothermal deposits in silicic-volcanic rocks (analcime, chabazite, clinoptilolite, erionite, heulandite, laumontite, mordenite, phillipsite, stilbite, and wairakite) calculated assuming an aqueous silica activity in equilibrium with quartz. Reproduced with the kind permission of the Mineralogical Society of America from [116], published in Reviews in Mineralogy and Petrology (2002).

different temperatures, one would need to refine crystal structures of all particular zeolites at particular temperatures, which cannot be achieved. However, it is possible to investigate evolution of **topological** information of zeolite frameworks, taking into account their complexity parameters reported in [114].

The topological complexity values for the zeolites under consideration are listed in Tab. 2.2. Here each zeolite is characterized by the three-letter symbol that corresponds to the particular topology of tetrahedral framework that it is based upon. It is important to note that clinoptilolite and heulandite are based upon the tetrahedral framework of the same type, HEU, whereas analcime and wairakite contain frameworks of the ANA topology. In order to distinguish between these minerals, we use superscript symbols, e.g. HEU^c and HEU^h correspond to clinoptilolite and heulandite, respectively, whereas ANA^a and ANA^w correspond to analcime and wairakite, respectively. Fig. 2.20 represents modified versions of the diagrams shown in Fig. 2.19, where the fields of mineral phases are replaced by the fields of the topological types of their frameworks.

To estimate the average topological complexity of zeolite mineral-parageneses at different temperatures, one may use the parameter called average atomic topological

Tab. 2.2: Framework types (FT) and topological complexity parameters for zeolites occurring in hydrothermal deposits in silicic-volcanic rocks.

Mineral name	Chemical formula	FT	I_G [bits/atom]	$I_{G,total}$ [bits/cell]
analcime	$Na[AlSi_2O_6](H_2O)$	ANA	0.918	66.117
chabazite	$(Ca_{0.5},Na,K)_4[Al_4Si_8O_{12}](H_2O)_{12}$	CHA	2.252	81.059
clinoptilolite	$(Na,K)_6[Al_6Si_{30}O_{72}](H_2O)_{20}$	HEU	3.866	208.764
erionite	$K_2(Ca_{0.5},Na)_8[Al_{10}Si_{26}O_{72}](H_2O)_{28}$	ERI	2.837	306.352
heulandite	$(Na,K)Ca_4[Al_9Si_{27}O_{72}](H_2O)_{24}$	HEU	3.866	208.764
laumontite	$Ca_4[A_{l8}Si_{16}O_{48}](H_2O)_{18}$	LAU	3.281	118.117
mordenite	$(Na_2,K_2,Ca)_4[Al_8Si_{40}O_{96}](H_2O)_{28}$	MOR	3.670	264.235
phillipsite	$K_2(Ca_{0.5},Na)_4[Al_6Si_{10}O_{32}](H_2O)_{12}$	PHI	2.918	140.078
stilbite	$NaCa_4[Al_9Si_{27}O_{72}](H_2O)_{30}$	STI	3.273	176.764
wairakite	$Ca_8[Al_{16}Si_{32}O_{96}](H_2O)_{16}$	ANA	0.918	66.117

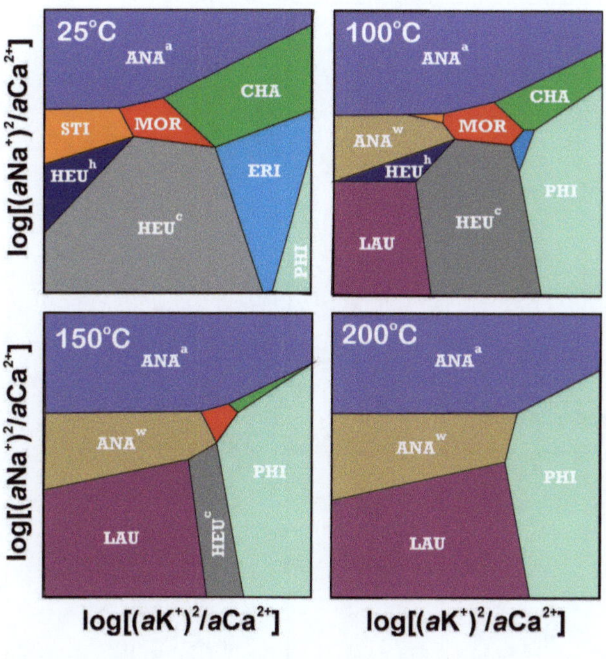

Fig. 2.20: Modified version of the diagram shown in Fig. 2.19, emphasizing topological types of zeolite frameworks occurring in particular mineral phases.

complexity, ^{at}ATC, for the section of a phase diagram defined as

$$^{at}\mathrm{ATC} = \sum_{i=1}^{n} [\delta_i \times {}^{i}I_G], \qquad (2.15)$$

where ${}^{i}I_G$ is a topological complexity of the ith phase measured as topological information per atom, n is the number of phases in the system, and δ_i is the normalized specific area of the ith phase in the diagram such that

$$\sum_{i=1}^{n} \delta_i = 1. \tag{2.16}$$

By analogy, one may define global ^{gl}ATC as

$$^{gl}\mathrm{ATC} = \sum_{i=1}^{n} [\delta_i \times {}^{i}I_{G,\mathrm{total}}], \tag{2.17}$$

where ${}^{i}I_{G,\mathrm{total}}$ is a topological complexity of the ith phase measured as topological information per unit-cell.

The behavior of the ^{at}ATC and ^{gl}ATC values *versus* temperature in the selected zeolite system is shown in Fig. 2.21. It is highly interesting that the temperature de-

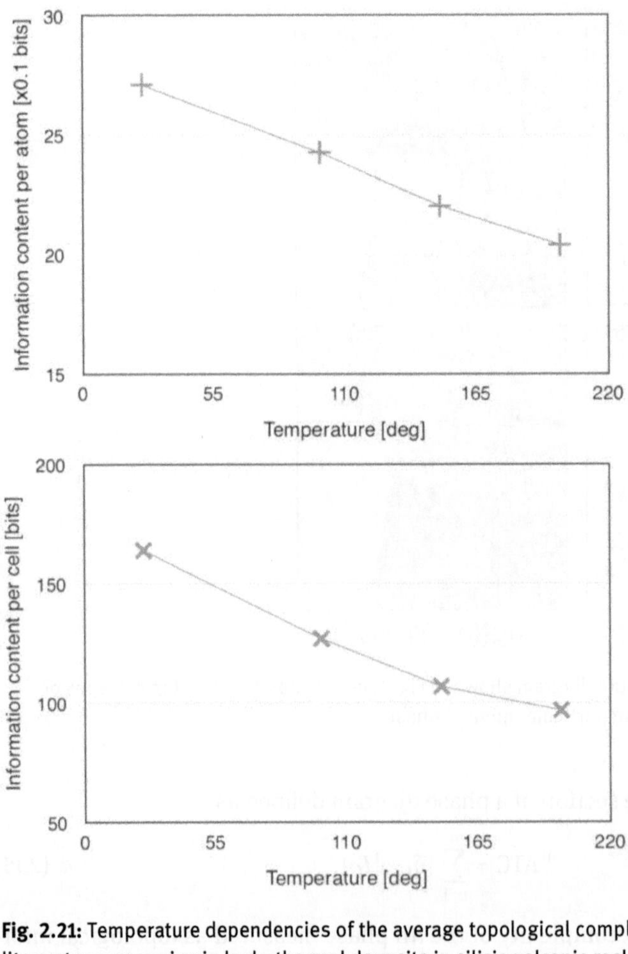

Fig. 2.21: Temperature dependencies of the average topological complexity parameters for the zeolite systems occurring in hydrothermal deposits in silicic-volcanic rocks.

pendencies of the average topological complexity parameters are almost **linear**: the topological complexity of the system decreases almost linearly with the increasing temperature.

The example of zeolite mineral parageneses considered supports the very general empirical rule that **structural and topological complexity of the chemically isolated system consisting of *n* crystalline phases decreases with the increase of temperature.** This rule reflects the fact that both configurational and vibrational entropies of crystalline materials are increasing with the increasing temperature. Though the exact relations between these two kinds of entropies and information-based complexity parameters are not yet clear, it is very probable that they both have positive correlation. Unfortunately, we are unaware of any systematic and widely applicable attempts to estimate configurational and vibrational entropies and their contributions to the total entropy of crystals.

The empirical rule formulated above can be violated, if there are other forces that govern phase transformations in crystals, e.g. chirality, magnetism, electricity, etc.

2.7 Macroevolution of structural complexity of minerals: Preliminary musings

The topic of global mineral evolution has emerged as a new paradigm in mineralogical sciences [3]. However, it is not always properly acknowledged that the concept of mineral evolution was first proposed and (to some extent) developed by Russian mineralogists in 1970–1980s [4, 117]. A short history of the concept of mineral evolution in Russian scientific literature was given by Krivovichev [15].

Hazen et al. [3] defined three eras and ten stages of mineral evolution in the Earth's history. The initial era (>4.55 Ga) is dominated by formation of minerals with a total information content that rarely exceeds 200 bits. At this stage, the appearance of species of intermediate complexity is usually associated with aqueous and thermal alteration events (formation of phyllosilicates and amphibole-supergroup minerals). The era of crust and mantle reworking (4.55 to 2.5 Ga) corresponds to the expansion of mineralogical diversity [3], but also results in accummulation of information due to the formation of a suite of minerals with intermediate complexity such as hornblende (146.877 bits/cell), tourmaline-group minerals (176–207 bits/cell), pollucite (103.637 bits/cell), beryl (132.825 bits/cell), scapolite (135.133 bits/cell), staurolite (163.329 bits/cell), cordierite (120.175 bits/cell), etc. The third era of mineral evolution was identified by Hazen et al. [3] as an era of bio-mediated mineralogy (2.5 Ga to present), when a diversity of new formation pathways of minerals was triggered by the activity of living organisms. This era is characterized by the appearance of minerals produced by biomineralization processes; some of these minerals are especially information-rich: vivianite $Fe_3(PO_4)_2(H_2O)_8$ (128.750 bits/cell), struvite

$Mg(NH_4)(PO_4)(H_2O)_6$ (245.763 bits/cell), schwertmannite $Fe_{16}O_{16}(OH)_{16}(H_2O)_2$ (282.193 bits/cell), hannayite $Mg_3(NH_4)_2(HPO_4)_4(H_2O)_8$ (301.775 bits/cell), whitlockite $Ca_{2.89}Mg_{0.11}(PO_4)_2$ (375.013 bits/cell), melanterite (517.528/cell), hazenite $KNaMg_2(PO_4)_2(H_2O)_{14}$ (936.313/cell), bakhchisaraitsevite $Na_2(Mg,Fe)_5(PO_4)_4(H_2O)_7$ (1128.771 bits/cell), etc. Extensive inorganic and bio-weathering activity during the Phanerozoic era is also responsible for the formation of very complex and information-rich minerals such as bouazzerite (3035.201 bits/cell), vandendriesscheite (2835.307 bits/cell), pertlikite $K_2(Fe,Mg)_2Mg_3Fe_3Al(SO_4)_{12}(H_2O)_{18}$ (1865.225 bits/cell), richetite $(Mg,Fe)Pb_9((UO_2)_{18}O_{18}(OH)_{12})_2(H_2O)_{41}$ (1748.848 bits/cell), pushcharovskite $Cu(AsO_3OH)(H_2O)$ (1640.967 bits/cell), voltaite $K_2Mg_5Fe_4(SO_4)_{12}(H_2O)_{18}$ (1566.017 bits/cell), etc. Low-temperature hydrothermal activity results in the appearance of very complex zeolites (paulingite, tschoertnerite, mutinaite) and secondary phosphates such as cacoxenite, $Al_4Fe_{21}(PO_4)_{17}O_6(OH)_{12}(H_2O)_{24}$ (1537.391 bits/cell). The general trend of mineral evolution corresponds to the accummulation of information both in terms of diversity and complexity of the mineral world. Whether this trend is related to the evolution of biological complexity [118] or it is a consequence of general evolution of complexity in complex systems is an open question.

2.8 Information storage and processing in minerals *versus* living matter

In order to estimate information capacity of inorganic crystal structures *versus* living matter, one has to keep in mind that, in living organisms, information is stored in the DNA molecule contained in each cell of the organism. The DNA contains a sequence of four nucleobases, A (adenine), T (thymine), C (cytosine), and G (guanine), which can be considered as letters of an alphabet. The DNA of a simplest organism such as *E. coli* bacteria contains about 4×10^6 nucleotides. If different bases occur in the DNA of *E. coli* equally often, then for each nucleotide its probability of occurrence is equal to 1/4, which means that each base bears exactly 2 bits of Shannon information. Therefore, the total Shannon information of the DNA molecule of *E. coli* is about 8×10^6 bits [119]. The largest human chromosome, chromosome number 1, contains about 220×10^6 nucleotides, which, under condition of equal probability, equals to 440×10^6 bits of Shannon information.

As it was mentioned above, the most complex inorganic crystal structure of $Al_{55.4}Cu_{5.4}Ta_{39.1}$ contains 48 538.637 bits of Shannon information per unit cell, whereas the most complex mineral, paulingite, registers 6 766.998 bits. Therefore, in very rough terms, Shannon information capacity of inorganic crystals and living organisms differs by about three orders of magnitude. To use a simple analogy, the **most complex** mineral corresponds to one page of a book, whereas the **simplest** organism can be compared to a 1000-page book.

The difference of information capacity between inorganic and biological worlds can be compared to the difference in their diversity. According to the recent estimates [120], there are ~1.2 million currently valid biological species, and the estimated number of eukaryotic species only is ~8.7±1.3 million. The number of mineral species known today does not exceed 6000, and there are no estimates as to how large this number can be. Therefore, biodiversity exceeds diversity of mineral species by about three orders of magnitude, which is comparable to the difference in their information storage capacity.

There are, of course, many other differences between inorganic and biological species in terms of their information storage and processing. The most important one is in the character of the information: it is functional in organisms and obviously non-functional in crystals.

2.9 Conclusions

The approach to the analysis of minerals, their composition, crystal structure and evolution, in terms of information theory developed in this chapter and other works [14–16, 114] may provide interesting and important insights into those basic features of mineral behavior that depend upon structural and topological organization of matter in crystals. It also provides a general framework for the analysis of large amount of empirical data on the crystal structures of minerals accumulated in the mineralogical literature over the last hundred years. Among the directions for the further development of the approach one may identify the following: (i) the behavior of average structural complexity in the course of global mineral evolution; (ii) the fluctuations of complexity in the local-scale mineralogical and geochemical processes and their relations to the changes in chemical and thermodynamic parameters of the environment; (iii) the relations between complexity (understood as information) and energy from the viewpoint of current trends in information-to-energy conversion studies [121, 122]; (iv) the relations between information-based complexity parameters and total entropy of minerals.

Acknowledgment: This work was supported by St. Petersburg State University (grant 3.38.136.2014).

Bibliography

[1] Pavlinov IYa. Introduction to Contemporary Phylogenetics (a Cladogenetic Aspect). Moscow, Russia, KMK Scientific Press LTd, 2005 (in Russian).

[2] Lehn JM. Perspectives in chemistry–steps towards complex matter. Angew Chem Int Ed 2013, 52, 2836–2850.

[3] Hazen RM, Papineau D, Bleeker W, Downs RT, Ferry JM, McCoy TJ, Sverjensky DA, Yang H. Mineral evolution. Amer Mineral 2008, 93, 1693–1720.
[4] Yushkin NP. Evolutionary ideas in modern mineralogy. Zapiski Vsesoyuznogo Mineralogicheskogo Obshchestva 1982, 116(4), 432–442 (in Russian).
[5] Avery J. Information Theory and Evolution, World Scientific Publishing, Singapore, 2003.
[6] Bonchev D, Rouvray DH, eds. Complexity in Chemistry, Biology, and Ecology. Springer, New York, 2005.
[7] Davies P, Gregersen N, eds. Information and The Nature of Reality. From Physics and Metaphysics. Cambridge University Press, Cambridge, U.K., 2010.
[8] Graham DJ. Chemical Thermodynamics and Information Theory with Applications. CRC Press, Taylor & Francis Group, Boca Raton, FL, USA, 2011.
[9] Downs RT. The RRUFF Project: an integrated study of the chemistry, crystallography, Raman and infrared spectroscopy of minerals. Program and Abstracts of the 19th General Meeting of the International Mineralogical Association in Kobe, Japan, 2006. O03–13.
[10] Buerger MJ, Venkatakrishnan V. Serendibite, a complicated, new, inorganic crystal structure. Proc Natl Acad Sci 1979, 71, 4348–4351.
[11] Moore PB, Shen J. Crystal structure of steenstrupine: A rod structure of unusual complexity. Tscherm Mineral Petrogr Mitt 1983, 31, 47–67.
[12] Burns PC, Hawthorne FC. Structure and hydrogen bonding in preobrazhenskite, a complex heteropolyhedral borate. Can Mineral 1994, 32, 387–396.
[13] Bindi L, Nestola F, Guastoni A, Secco L. The crystal structure of dalnegroite, $Tl_{5-x}Pb_{2x}(As,Sb)_{21-x}S_{34}$: a masterpiece of structural complexity. Mineral Mag 2010, 74, 999–1012.
[14] Krivovichev SV. Topological complexity of crystal structures: quantitative approach. Acta Crystallogr 2012, A68, 393–398.
[15] Krivovichev SV. Structural complexity of minerals: information storage and processing in the mineral world. Miner Mag 2013, 77, 275–326.
[16] Krivovichev SV. Which inorganic structures are the most complex? Angew Chem Int Ed 2014, 53, 654–661.
[17] Pauling L. The principles determining the structure of complex ionic crystals. J Amer Chem Soc 1929, 51, 1010–1026.
[18] Baur WH, Tillmanns ET, Hofmeister W. Topological analysis of crystal structures. Acta Crystallogr 1983, B39, 669–674.
[19] Oganov AR, Valle M. How to quantify energy landscapes of solids. J Chem Phys 2009, 130, 104504.
[20] Lister SE, Radoslavljevic Evans I, Howard JAK, Coelho A, Evans JSO. $Mo_2P_4O_{15}$ – the most complex oxide structure solved by single crystal methods? Chem Commun 2004, 2540–2451.
[21] Lister SE, Radoslavljevic Evans I, Evans JSO. Complex Superstructures of $Mo_2P_4O_{15}$. Inorg Chem 2009, 48, 9271–9281.
[22] Mackay AL, Klinowski J. Towards a grammar of inorganic structure. Comp Math Appl 1986, B12, 803–824.
[23] Shannon E, Weaver W. The Mathematical Theory of Communications. University of Illinois Press, Urbana, IL, 1949.
[24] Moore PB. Structural hierarchies among minerals containing octahedrally coordinated oxygen. II. Systematic retrieval and classification of octahedral edge-sharing clusters: an epistemiological approach. N Jb Mineral Abh 1974, 120, 205–227.
[25] Hawthorne FC. Graphical enumeration of polyhedral clusters. Acta Crystallogr 1983, A39, 724–736.
[26] Rashevsky N. Life, information theory, and topology. Bull Math Biophys 1955, 17, 229–235.

[27] Trucco E. A note on the information content of graphs. Bull Math Biophys 1956, 18, 129–135.
[28] Dehmer M, Mowshowitz A. A history of graph entropy measures. Inf Sci 2011, 181, 57–78.
[29] Chung SJ, Hahn Th, Klee WE. Nomenclature and generation of three-periodic nets: the vector method. Acta Crystallogr 1984, A40, 42–50.
[30] Klee WE. Crystallographic nets and their quotient graphs. Cryst Res Technol 2004, 39, 959–968.
[31] Bandy MC. Lopezite, a new mineral. Amer Mineral 1937, 22, 929–930.
[32] Weakley TJR, Ylvisaker ER, Yager RJ, Stephens JE, Wiegel RD, Mengis M, Wu P, Photinos P, Abrahams SC. Phase transitions in $K_2Cr_2O_7$ and structural redetermination of phase II. Acta Crystallogr 2004, B60, 705–715.
[33] Steurer W. Measures of complexity. Acta Crystallogr 2011, A67, C184.
[34] Krivovichev SV, Kir'yanova EV, Filatov SK, Burns PC. β-$K_2Cr_2O_7$. Acta Crystallogr 2000, C56, 629–630.
[35] Guo GC, Cheng WD, Chen JT, Zhuang HH, Huang JS, Zhang QE. Monoclinic $Mg_2B_2O_5$. Acta Crystallogr 1995, C51, 2469–2471.
[36] Pant AK. A reconsideration of the crystal structure of β-$Na_2Si_2O_5$. Acta Crystallogr 1968, B24, 1077–1083.
[37] Mitchell M. Complexity. A Guided Tour. Oxford University Press, Oxford, New York, 2011.
[38] Ilachinski A. Cellular Automata. A Discrete Universe. World Scientific Publishing Co., New Jersey, London, Singapore, Hong Kong, 2001.
[39] Estevez-Rams E., González-Férez R. On the concept of long-range order in solids: the use of algorithmic complexity. Z Kristallogr 2009, 224, 179–184.
[40] Mackay AL. Crystal symmetry. Phys Bull 1976, 1976, 495–496.
[41] Krivovichev SV. Crystal structures and cellular automata. Acta Crystallogr 2004, A60, 257–262.
[42] Shevchenko VYa, Krivovichev SV. Where are genes in paulingite? Mathematical principles of formation of inorganic materials on the atomic level. Struct Chem 2008, 19, 571–577.
[43] Krivovichev SV. Actinyl compounds with hexavalent elements (S, Cr, Se, Mo) – Structural diversity, nanoscale chemistry, and cellular automata modeling. Eur J Inorg Chem 2010, 2594–2603.
[44] Krivovichev SV. Algorithmic crystal chemistry: a cellular automata approach. Crystallogr Rep 2012, 57, 10–17.
[45] Shevchenko VYa, Krivovichev SV, Mackay A. Cellular automata and local order in the structural chemistry of the lovozerite group minerals. Glass Phys Chem 2010, 36, 1–9.
[46] Krivovichev SV, Shcherbakova EP, Nishanbaev TP. Crystal structure of svyatoslavite and evolution of complexity during crystallization of the $CaAl_2Si_2O_8$ melt: a structural automata description. Can Mineral 2012, 50, 585–592.
[47] Krivovichev SV. On the algorithmic complexity of crystals. Miner Mag 2014, 78, 415–435.
[48] Delgado Friedrichs O, O'Keeffe M, Yaghi OM. Three-periodic nets and tilings: regular and quasiregular nets. Acta Crystallogr 2003, A59, 22–27.
[49] Hopcroft JE, Motwani R, Ullman JD. Introduction to Automata Theory, Languages and Computation. Addison-Wesley, Boston, 2001.
[50] Morey J, Sedig K, Mercer RE, Wilson W. Crystal lattice automata. In: Watson BW, Wood D, eds. CIAA 2001, LNCS 2494, Springer-Verlag, Berlin-Heidelberg, 2002, 214–220.
[51] Oberhagemann U, Bayat P, Marler B, Gies H, Rius J. A layer silicate: synthesis and structure of the zeolite precursor RUB-15 – $[N(CH_3)_4]_8[Si_{24}O_{52}(OH)_4]\cdot 20H_2O$. Angew Chem Int Ed 1996, 35, 2869–2871.
[52] Blatov VA, Shevchenko AP, Serezhkin VN. TOPOS 3.2 – a new version of the program package for multipurpose crystal-chemical analysis. J Appl Crystallogr 2001, 33, 1193.

[53] Warren BE, Modell DI. The structure of vesuvianite $Ca_{10}Al_4(Mg,Fe)_2Si_9O_{34}(OH)_4$. Z Kristallogr 1931, 78, 422–432.
[54] Allen FM, Burnham CW. A comprehensive structure-model for vesuvianite: Symmetry variation and crystal growth. Can Mineral 1992, 30, 1–18.
[55] Groat LA, Hawthorne FC, Ercit TS. The role of fluorine in vesuvianite: a crystal-structure study. Can Mineral 1992, 30, 1065–1075.
[56] Gnos E, Armbruster T. Relationship among metamorphic grade, vesuvianite rod polytypism, and vesuvianite composition. Amer Mineral 2006, 91, 862–870.
[57] Moore PB. Structural hierarchies among minerals containing octahedrally coordinating oxygen. I. Stereoisomerism among corner-sharing octahedral and tetrahedral chains. N Jb Mineral Mh 1970, 163–173.
[58] Machatschki F. Zur Frage der Struktur und Konstitution der Felspate. (Gleichzeitig vorläufige Mitteilung über die Prinzipien des Baues der Silikate. Centr Mineral Geol Paläontol A, Mineral Petrogr 1928, 1928, 97–104.
[59] Bragg WL. The structure of silicates. Z Kristallogr 1930, 74, 237–305.
[60] Hawthorne FC. The structure hierarchy hypothesis. Mineral Mag 2014, 78, 957–1027.
[61] Liebau F. Structural Chemistry of Silicates. Structure, Bonding and Classification. Springer-Verlag, Berlin, 1985.
[62] Salthe SN. Hierarchical structures. Axiomathes 2012, 22, 355–383.
[63] Strunz H, Nickel E. Strunz Mineralogical Tables. Chemical-Structural Mineral Classification System. 9th ed. Schweizerbart Science Publishers, Stuttgart, 2001.
[64] Simon HA. The architecture of complexity. Proc Amer Philos Soc 1962, 106, 467–482.
[65] Ferraris G, Makovicky E, Merlino S. Crystallography of Modular Materials. Oxford University Press, Oxford, 2004.
[66] Goldsmith JR. A "simplexity principle" and its relation to "ease" of crystallization. J Geol 1953, 61, 439–451.
[67] Morse JW, Casey WH. Ostwald processes and mineral paragenesis in sediments. Amer J Sci 1988, 288, 537–5560.
[68] Navrotsky A. Energetic clues to pathways to biomineralization: precursors, clusters, and nanoparticles. Proc Natl Acad Sci USA 2004, 101, 12096–12101.
[69] Darling RS, Chou IM, Bodnar RJ. An occurrence of metastable cristobalite in high-pressure garnet granulite. Science 1997, 276, 91–93.
[70] Okamoto A, Saishu H, Hirano N, Tsuchiya N. Mineralogical and textural variation of silica minerals in hydrothermal flow-through experiments: implications for quartz vein formation. Geochim Cosmochim Acta 2010, 74, 3692–3706.
[71] Herdianita NR, Browne PRL, Rodgers KA, Campbell KA. Mineralogical and textural changes accompanying ageing of silica sinter. Mineral Dep 2000, 35, 48–62.
[72] Chesnokov BV, Lotova EV, Pavlyuchenko VS, Nigmatulina EN, Usova LV, Bushmakin AF, Nishanbaev TP. Svyatoslavite $CaAl_2Si_2O_8$ (orthorhombic) – a new mineral. Zapiski Vsesoyuznogo Mineralogicheskogo Obshchestva 1989, 118(2), 111–114 (in Russian).
[73] Chesnokov BV, Lotova EV, Nigmatulina EN, Pavlyuchenko VS, Bushmakin AF. Dmishteinbergite $CaAl_2Si_2O_8$ (hexagonal) – a new mineral. Zapiski Vsesoyuznogo Mineralogicheskogo Obshchestva 1990, 119(3), 43–46 (in Russian).
[74] Davis GL, Tuttle OF. Two new crystalline phases of the anorthite compositon, $CaO{\cdot}Al_2O_3{\cdot}2SiO_2$. Amer J Sci 1952, Bowen Volume, 107–114.
[75] Abe T, Tsukamoto K, Sunagawa I. Nucleation, growth and stability of $CaAl_2Si_2O_8$ polymorphs. Phys Chem Miner 1991, 17, 473–484.
[76] Daniel I, Gillet P, McMillan PF, Richet P. An in-situ high-temperature structural study of stable and metastable $CaAl_2Si_2O_8$ polymorphs. Mineral Mag 1995, 59, 25–33.

[77] Abe T, Sunagawa I. Hexagonal $CaAl_2Si_2O_8$ in a high temperature solution; metastable crystallization and transformation to anorthite. Mineral J 1995, 17, 257–281.
[78] Takéuchi Y, Donnay G. The crystal structure of hexagonal $CaAl_2Si_2O_8$. Acta Crystallogr 1959, 12, 465–470.
[79] Dimitrijevic R, Dondur V, Kremenovic A. Thermally induced phase transformations of Ca-exchanged LTA and FAU zeolite frameworks: Rietveld refinement of the hexagonal $CaAl_2Si_2O_8$ diphyllosilicate structure. Zeolites 1999, 16, 294–300.
[80] Nestola F, Mittempergher S, Di Toro G, Zorzi F, Pedron D. Evidence of dmisteinbergite (hexagonal form of $CaAl_2Si_2O_8$) in pseudotachylyte: a tool to constrain the thermal history of a seismic event. Amer Mineral 2010, 95, 405–409.
[81] Ma C, Krot AN, Bizzarro M. Discovery of dmisteinbergite (hexagonal $CaAl_2Si_2O_8$) in the Allende meteorite: a new member of refractory silicates formed in the Solar nebula. Amer Mineral 2013, 98, 1368–1371.
[82] Hwang SL, Shen P, Chu HT, Yui TF, Liou JG, Sobolev NV. Kumdykolite, an orthorhombic polymorph of albite, from Kokchetav ultrahigh-pressure massif, Kazakhstan. Eur J Mineral 2010, 21, 1325–1334.
[83] Neméth P, Lehner SW, Petaev MI, Buseck PR. Kumdykolite, a high-temperature feldspar from an enstatite chondrite. Amer Mineral 2013, 98, 1070–1073.
[84] Hwang SL, Shen P, Chu HT, Yui TF, Liou JG, Sobolev NV, Zhang RY, Shatsky VS, Zayachkovsky AA. Kokchetavite: a new potassium-feldspar polymorph from the Kokchetav ultrahigh-pressure terrane. Contr Mineral Petrol 2004, 148, 380–389.
[85] Fasshauer DW, Chatterjee ND, Marler B. Synthesis, structure, thermodynamic properties and stability relations of K-cymrite, $K[AlSi_3O_8]\cdot H_2O$. Phys Chem Miner 1997, 24, 455–462.
[86] Graham CM, Tareen JAK, McMillan PF, Lowe BM. An experimental and thermodynamic study of cymrite and celsian stability in the system $BaO–Al_2O_3–SiO_2–H_2O$. Eur J Mineral 1992, 4, 251–269.
[87] Hyatt MJ, Bansal NP. Crystal growth kinetics in $BaOAl_2O_32SiO_2$ and $SrOAl_2O_32SiO_2$ glasses. J Mater Sci 1996, 31, 172–184.
[88] Viswanathan K, Harneit O, Epple M. Hydrated barium aluminosilicates, $BaAl_2Si_2O_8H_2O$, and their relations to cymrite and hexacelsian. Eur J Mineral 1992, 4, 271–278.
[89] Evans BW, Johannes W, Oterdoom H, Trommsdorff V. Stability of chrysotile and antigorite in the serpentinite multisystem. Schweiz Mineral Petrogr Mitt 1976, 56, 79–93.
[90] Evans BW. The serpentinite system revisited: chrysotile is metastable. Int Geol Rev 2004, 46, 479–506.
[91] Wunder B, Wirth R, Gottschalk M. Antigorite: pressure and temperature dependence of polysomatism and water content. Eur J Mineral 2001, 13, 485–495.
[92] Dungan MA. Metastability in serpentine-olivine equilibria. Amer Mineral 1979, 62, 1018–1029.
[93] Uehara S. TEM and XRD study of antigorite superstructures. Can Mineral 1998, 36, 1595–1605.
[94] Dodony I, Posfai M, Buseck PR. Revised structure models for antigorite: an HRTEM study. Amer Mineral 2002, 87, 1443–1457.
[95] Capitani G, Mellini M. The modulated crystal structure of antigorite: the m = 17 polysome. Amer Mineral 2004, 89, 147–158.
[96] Capitani GC, Mellini M. The crystal structure of a second antigorite polysome (m = 16), by single-crystal synchrotron diffraction. Amer Mineral 2006, 91, 394–399.
[97] Mellini M, Trommsdorff V, Compagnoni R. Antigorite polysomatism: behaviour during progressive metamorphism. Contr Mineral Petrol 1987, 97, 147–155.

[98] Velde B. Experimental determination of muscovite polymorph stabilities. Amer Mineral 1965, 50, 436–449.
[99] Mukhamet-Galeyev AP, Pokrovskiy VA, Zotov AV, Ivanov IP, Samotoin N. Kinetics and mechanism of hydrothermal crystallization of $2M_l$ muscovite: An experimental study. Int Geol Rev 1985, 27, 1352–1364.
[100] Meunier A, Velde BD. Illite: Origin, Evolution and Metamorphism. Springer-Verlag, Berlin Heidelberg New York, 2004.
[101] Baxter Grubb SM, Peacor DR, Jiang WT. Transmission electron microscope observations of illite polymorphism. Clays and Clay Minerals 1991, 39, 540–550.
[102] Navrotsky A, Trofimlyuk O, Levchenko AA. Thermochemistry of microporous and mesoporous materials. Chem Rev 2009, 109, 3885–3902.
[103] Krivovichev SV, Siidra OI, Nazarchuk EV, Burns PC, Depmeier W. Particular topological complexity of lead oxide blocks in $Pb_{31}O_{22}X_{18}$ (X = Br, Cl). Inorg Chem 2006, 45, 3846–3848.
[104] Aliev A, Kovrugin VM, Colmont M, Terryn C, Huve M, Siidra OI, Krivovichev SV, Mentre O. Revised bismuth chloroselenite system: evidence of a noncentrosymmmetric structure with a giant unit cell. Cryst Growth Des 2014, 3026–3034.
[105] Moore PB. Pegmatite phosphates: descriptive mineralogy and crystal chemistry. Mineral Rec 1973, 4, 103–130.
[106] Hawthorne FC. Structure and chemistry of phosphate minerals. Mineral Mag 1998, 62, 141–164.
[107] Wang A, Ling ZC. Ferric sulfates on Mars: a combined mission data analysis of salty soils at Gusev crater and laboratory experimental investigations. J Geophys Res Planets 2011, 116 (E0), 22.
[108] Xu W, Tosca NJ, McLennan SM, Parise JB. Humidity-induced phase transitions of ferric sulfate minerals studied by in situ and ex situ X-ray diffraction. Amer Mineral 2009, 94, 1629–1637.
[109] Ling ZC, Wang A. A systematic spectroscopic study of eight hydrous ferric sulfates relevant to Mars. Icarus 2010, 209, 422–433.
[110] Wang A, Ling ZC, Freeman JJ, Kong W. Stability field and phase transition pathways of hydrous ferric sulfates in the temperature range 50 °C to 5 °C: implication for martian ferric sulfates. Icarus 2012, 218, 622–643.
[111] Jambor JL, Nordstrom DK, Alpers CN. Metal-sulfate salts from sulfide mineral oxidation. Rev Mineral Geochem 2000, 40, 303–350.
[112] Hawthorne FC, Krivovichev SV, Burns PC. The crystal chemistry of sulfate minerals. Rev Mineral Geochem 2000, 40, 1–112.
[113] Christy AG, Mills SJ, Kampf AR, Housley RM, Thorne B, Marty J. The relationship between mineral composition, crystal structure and paragenetic sequence: the case of secondary Te mineralization at Bird Nest drift, Otto Mountain, California, USA. Mineral Mag 2015, in press.
[114] Krivovichev SV. Structural and topological complexity of zeolites: an information-based approach. Micropor Mesopor Mater 2013, 171, 223–229.
[115] Bargar KE, Keith TEC. Calcium zeolites in rhyolitic drill cores from Yellowstone National Park, Wyoming. In: Ming DW, Mumpton FA, eds. Natural Zeolites '93: Occurrence, Properties, Use. Intl Comm Nat Zeol, Brockport, New York, 1999, 69–86.
[116] Chipera SJ, Apps JA (2001): Geochemical stability of natural zeolites. Reviews in Mineralogy and Geochemistry, 45, 117–161.
[117] Zhabin AG. Is there evolution of mineral species on Earth? Dokl Earth Sci 1981, 247, 142–144.
[118] Adami C. What is complexity? BioEssays 2002, 24, 1085–1094.
[119] Bradley WL. Information, entropy and the origin of life. In: Dembski WA, Ruse M, eds. Debating Design. From Darwin to DNA. New York, USA, Cambridge University Press, 2006, 331–351.

[120] Mora C, Tittensor DP, Adl S, Simpson AGB, Worm B. How many species are there on Earth and in the ocean? PLOS Biology 2011, 9, e1001127.
[121] Toyabe S, Sagawa T, Ueda M, Muneyuki E, Sano M. Experimental demonstration of information-to-energy conversion and validation of the generalized Jarzynski equality. Nature Phys 2010, 6, 988–992.
[122] Bérut A, Arakelyan A, Petrosyan A, Ciliberto S, Dillenschneider R, Lutz E. Experimental verification of Landauer's principle linking information and thermodynamics. Nature 2012, 483, 187–190.

Sergey V. Churakov

3 Ab initio simulations of mineral surfaces: Recent advances in numerical methods and selected applications

3.1 Introduction

Traditional mineralogy and crystallography focus on the characterization of the structure of bulk minerals. Variations of structural parameters and physical properties of a mineral can be related to changes in chemical composition. For a given structural polymorph the variations of lattice parameters with phase composition are comparatively small. In contrast, a large diversity of crystal facets can be identified in mineral crystals having virtually identical bulk compositions. Crystal habit is a record of the history and physicochemical conditions of mineral growth. Small variation in growth conditions, such as cooling rate, degree of saturation and the presence of minor elements may have substantial consequences for the crystal morphology.

The formation of minerals and the stability of mineral assemblages in rocks can be described by thermodynamic equilibria between bulk phases. The rates of mineral dissolution and growth are controlled by the surface reactivity. Since the composition of a mineral surface can differ substantially from the bulk, surface layers may inhibit or catalyze dissolution or precipitation processes. Fundamental understanding the structure, reactivity and composition of mineral surfaces is essential for model-based description of geochemical processes.

Before the advent of surface sensitive techniques, the concepts of surface structures were derived based on the knowledge of bulk crystal structures, empirical systematics of crystal morphologies and basic electrostatic arguments. The development of surface sensitive diffraction methods (e.g. low energy electron diffraction, grazing incidence X-ray diffraction), electron spectroscopy methods (e.g. Auger electron spectroscopy, electron energy loss spectroscopy, photo electron spectroscopy) and high-resolution surface microscopy techniques (atomic force microscopy and scanning tunneling microscopy) has revolutionized our view of mineral surfaces. The surface analysis has revealed that structure of near surface atomic layers may undergo substantial geometrical transformations different to the symmetry of the bulk phase. These structural changes lead to modifications of the electronic structure in the near surface layers which are responsible for a number of surface related phenomena such as the narrowing of the band gap and enhancement or reduction of electron conductivity, etc. Finally, a stable surface composition and structure depend on the external chemical conditions. Mineral surfaces are often covered with a passivating layer which may inhibit both dissolution and growth processes.

Interpretation of experimental data is seldom straightforward. It always relies on a model and several different models can be consistent with the measurements. Atomistic simulations may help to bound the set of possible models or model parameters and to provide complementary information about the system which is hard to access experimentally. Atomistic simulations have been widely used to predict the physical properties and structure of solid phases. At the beginning of the computer modeling era, simulations with empirical interaction potentials were particularly successful thanks to the low computational costs. The parameters for empirical interaction potentials are calibrated against known structures and their physical properties. The predictions with empirical potentials are highly reliable as long as the short-range coordination of atoms and molecules in the studied system is similar to the one used for the calibration. This is generally not valid for surfaces. An additional complication arises due to surface reactivity. In most cases empirical force fields are not capable of handling changes in chemical bonding. For these reasons, mineral surfaces are preferably investigated using quantum mechanical calculations.

Quantum mechanical modelling of mineral surfaces is a relatively new and rapidly evolving field. The aim of this review is to give a short introduction to modern quantum mechanical simulation techniques and to illustrate their application to several intriguing phenomena at mineral surfaces. Obviously the scope of surface science is very broad. Several text books [1, 2] and reviews [3, 4] from the field of solid state physics cover a wide range of surface phenomena. Advances in atomistic simulations of carbonate, sulfate and selected silicate minerals, have been reviewed recently in a broader methodological context [5]. Unlike other reviews, this paper is focused on surface phenomena rather than on particular mineral systems. The topics reviewed in detail are surface reconstructions, changes in electronic properties of the surface (with respect to bulk), chemical reactivity, acid-base equilibria and redox processes. In all of these phenomena substantial changes in chemical bonding at the surface takes place and thus the use of electronic structure calculations is indispensable.

The review is organized as follow. The most important structural phenomena and commonly used terminology are introduced in Section 3.2. State-of-the-art atomistic simulation approaches are reviewed with a strong emphasis on quantum mechanical electronic structure calculations in Section 3.3. Case studies illustrating application of the theoretical methods are considered in Section 3.4. Finally, perspectives for future research are discussed.

3.2 Theoretical background and simulation methods

3.2.1 Structural rearrangements and surface symmetry

In a bulk crystal each atom experiences forces from neighboring atoms. At the position of equilibrium the total force acting on an atom is zero. If a crystal is cleaved to form a surface, the force balance is disturbed. In order to equalize the forces and minimize the surface energy, atoms in the near surface layer and the electron density undergo structural distortions. The structure of the surface layers is rearranged relative to the bulk by means of *relaxation, rumpling* and *reconstruction.*

The relaxation is a change in the spacing between surface layers of a mineral compared to the bulk structure. Approximating ionic structures by a system of point charges with repulsive cores, a summation of the electrostatic forces between surface layers suggest an inward relaxation of the surface. For covalent structures the relaxation is generally also inward. This behavior can be explained on the basis of valence-bond theory. The shortening of the interlayer distance is thus a compensation for the decrease in the coordination of the surface atoms. Despite the fact that the ionic and covalent structures show the same relaxation trend, the behavior of ion-covalent structures is not straightforward and depends on the balance between the electrostatic and covalent contributions to the interaction energy.

Surface rumpling occurs on the surface of minerals composed of more than one type of atoms and result in an off-plane relative displacement of atoms in the near surface planes. Rumpling is weak in dense ionic structures, and becomes more pronounced as the covalent character of the chemical bonding increases. Rumpling can be explained by differences in the polarization of anions and cations. The presence of the surface disturbs the symmetry of the electrostatic potential and causes polarization of the electron density. Since the anions are more polarizable then cations, the induced dipole moment is stronger at anionic sites. The total force acting on atoms is towards the bulk, but is weaker for anions. Thus anions are displaced less than the cations.

Finally, the atoms in the surface layer can also be displaced within the plane coplanar to the surface. This displacement changes the two-dimensional symmetry and the dimensions of the surface cell with respect to the bulk lattice, and is referred to as surface reconstruction. In the most general case, surface reconstruction can be expressed by a 2 × 2 matrix which relates the surface lattice vectors of the unreconstructed lattice (a_S, b_S) to the translation vectors of the reconstructed one (a'_s, b'_s):

$$\begin{pmatrix} a'_s \\ b'_s \end{pmatrix} = \begin{pmatrix} m_{11} & m_{12} \\ m_{21} & m_{22} \end{pmatrix} \begin{pmatrix} a_s \\ b_s \end{pmatrix} \tag{3.1}$$

If the angle between the lattice vectors of the original and the reconstructed lattice does not change, a more intuitive notation proposed by Wood can be applied [6]:

$$(hkl) - m \times n - R\phi \tag{3.2}$$

where (hkl) are the Miller indexes of the surface; m and n are the scaling factors of the translation vectors of the reconstructed lattice with respect to the bulk symmetry; ϕ is the rotational angle needed to align the reconstructed and original lattice vectors. The factor $R\phi$is omitted if the angle is zero. If the superstructure is induced by adsorption

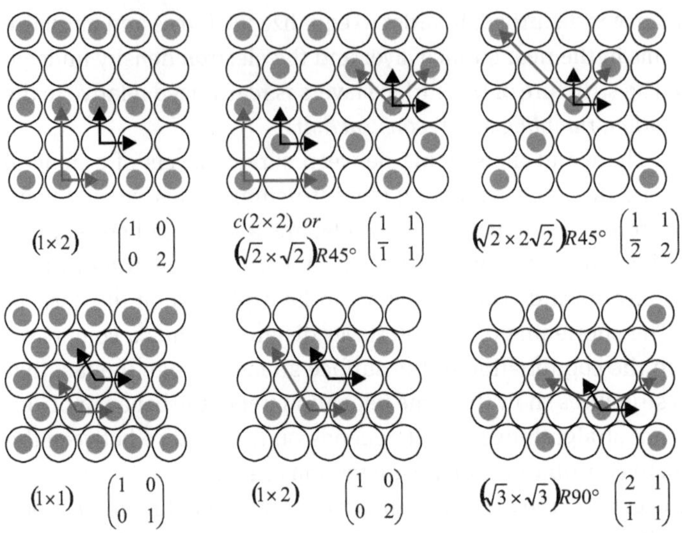

Fig. 3.1: Some example of Wood's and matrix notations of surface reconstructions for rectangular and hexagonal lattices. The lattice vectors of unreconstructed and reconstructed surfaces are shown as black and red arrows, respectively. Note that notations $c(2 \times 2)$ and $(\sqrt{2} \times \sqrt{2})$ R45° describe the same reconstruction.

the number of adsorbed molecules per supercell and type of adsorbate can be specified by its chemical formula at the end of the expression. For centered superstructures the symbol c is placed before the Miller index of the surface. Primitive cell can be indicated by the symbol p, but is omitted usually. Several examples of Wood's and matrix notation are illustrated in Fig. 3.1.

3.2.2 Stability of surfaces of simple ionic solids

The chemical bonding in most common minerals (e.g. silicates, oxides, sulfides, carbonates, etc.) can be well described as ionic-covalent interactions. This means that a significant redistribution of the electron density from the less electronegative atoms to the more electronegative atoms takes place, and the structure can be considered as a 3D arrangement of positively charged cations and negatively charged anions. Depending on the structure and the orientation of the lattice plane, three basic surface types can be imagined as illustrated in Fig. 3.2.

In type I surfaces, each atomic plane parallel to the surface is charge neutral and the surface has a zero dipole moment. In type II surfaces, atomic planes are not charge neutral, but the neighboring planes can be combined in charge neutral stoichiometric layers with a zero dipole moment. Finally, type III surfaces are formed by the alternation of negatively and positively charged planes which cannot be combined into units with zero dipole moment. According to basic electrostatic considerations, the surfaces

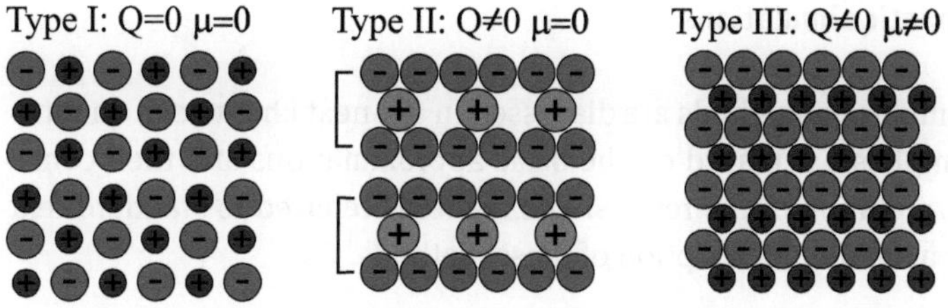

Fig. 3.2: Three types of ionic and partially ionic materials after Tasker [124]. Q is surface layer charge and μ is dipole moment.

of type I and II should have moderate surface energies and can be stable. In contrast, the electrostatic energy of a crystal with the type III surface diverges (to infinity). Thus type III surfaces can only be stabilized by substantial reconstruction within the surface layer.

The classical electrostatic arguments offer a good qualitative description of surface stability. However, numerous experimental and theoretical evidences exist that type III surfaces may be stable under certain conditions. The simple ionic model neglects the polarization of electron density at the interface, which can be sufficient to stabilize polar surfaces. For this reason electronic structure based methods and polarizable force-field models are superior to the non-polarizable force fields, and have to be used for the surface studies.

3.2.3 Influence of the surface on the electronic structure

According to the concept of band theory, electrons in solids can occupy certain energy levels (electronic bands) separated by forbidden energy regions (band gaps). These energy levels can be interpreted as the superposition of electron states in the reciprocal space and can be measured by spectroscopic methods or estimated by electronic structure calculations. In an insulator (the majority of minerals), the electron bands are either fully occupied by electrons (lower energy levels) or empty (higher energy levels). Conductors (metals) possess partially occupied energy levels (conduction bands). In half-metals, the energy bands of one set of electron spins have properties of an insulator whereas those of another set of electron spins are conductor like. The presence of a surface can modify the band structure in various ways. Atomic relaxations can shift energy levels of valence and conduction bands leading to a narrowing or broadening of the band gap. The magnitude of the band gap determines the acid-base properties of the surface and its electron affinity. Non-stoichiometric surface reconstruction, surface defects and sorption result in the formation of surface electronic states, which can change the electronic structure of a surface layer form a conductor to an insulator or vice versa.

3.2.4 Theory of atomistic simulations

Several molecular simulation methods are discussed in the next chapters at an introductory level. The emphasis is placed on the basic approximations and their consequences for the interpretation of the results. The reader is referred to standard textbooks, e.g. [7–11] for a detailed description of these methods.

3.2.4.1 Molecular dynamics and Monte Carlo simulation methods

In the Molecular Dynamics (MD) method the thermal motion of N particles enclosed in a volume V, e.g. atoms in solid, fluids or at material interface, can be modelled by solving the system of Newton's equations:

$$M_k \frac{d^2 \mathbf{R_k}}{dt^2} = \mathbf{F_k}[\{R_I\}] = -\frac{\partial U[\{\mathbf{R}_I\}]}{\partial \mathbf{R_k}}, \qquad k, I \in [1..N] \tag{3.3}$$

where M_k and $\mathbf{R_k}$ are mass and position of the *k-th* atom in the system, respectively; t is the time and U is the total interaction energy of the system, which in general depend on positions of all atoms in the system $\{\mathbf{R_I}\}$; $\mathbf{F_k}$ is the total force acting on the k-*th* atom. Apart from a few special cases, the eq. (3.3) has to be solved numerically. Obtained time series of atomic positions form a sampling set of a so called thermodynamics ensemble and the physical properties of the material can be calculated based on formulas of statistical mechanics. Alternatively, a representative sampling set can be obtained using a stochastic process (Monte-Carlo simulations method (MC)). The randomly generated atomic configurations are accepted or rejected depending on the change in the system energy. If the thermal motion is irrelevant, the equilibrium structure of a material at zero Kelvin can be obtained setting the left hand side of eq. (3.3) to zero and finding the atomic configuration which minimizes the energy of the system (geometry optimization).

Bulk phases contain as many as $\sim N^{23}$ atoms, which is not possible to handle in explicit numerical simulations. However, elementary chemical reactions and physical collision processes typically involve only few atoms in their direct neighborhood. The rest of the system (confinement) is seen by the reacting species as a mean field (the assumption of interaction locality is not appropriate under thermodynamic conditions close to second-order phase transitions). Thus, the mechanisms of complex chemical phenomena can be studied using comparatively small sets of atoms. Periodic boundary conditions are typically applied to simulate bulk materials. In this approach a "small" number of atoms, which can vary between 10^2–10^6 particles depending on the application, is modelled explicitly. The interaction with the rest of the system is approximated by the interaction with the periodic images of the central supercell. In solid state applications the simulations box typically span several unit cells and referred to as a supercell. Periodic boundary conditions can also be applied to "non-periodic" systems such as surfaces and interfaces. Care must be taken to use a sufficiently large simulation supercell in order to reduce artificial self-interaction between periodic images of the central supercell. If required, special methods can be applied to decouple the interaction between periodic images explicitly [12, 13].

To perform geometry optimization, MD (eq. 3.3) or MC simulations, the potential energy of the system U has to be known for each instantaneous atomic configuration. The model used to calculate the system energy determines by far the tractable size of the system, accuracy and computational costs of the simulations. In general, one can use empirical interaction parameters (empirical force field method) or solve the stationary Schrödinger equation for the electrons in the system.

3.2.4.2 Empirical force field method

In the empirical Force Field (FF) method the interaction energy is defined as the sum of various inter- and intra-molecular contributions, e.g.:

$$U = U_{\text{inter}}^{\text{Coulomb}} + U_{\text{inter}}^{\text{Repulsion}} + U_{\text{inter}}^{\text{Dispersion}} + U_{\text{inter}}^{\text{Many-Body}} + U_{\text{inter}}^{\text{Polarization}} + U_{\text{intra}} \tag{3.4}$$

The first three terms in eq. (3.4) account for pair-wise intermolecular interactions (Coulomb Repulsion, Dispersion). Various terms of increasing complexity can be included to account for many-body interaction and the polarization of the media. Intra- (within a molecule) and inter-molecular (between molecules) interactions can be distinguished for molecular systems.

The Coulomb energy is a sum of pair-wise interactions of point charges:

$$U^{\text{coulomb}} = \sum_{i>j} \frac{e^2 q_i q_j}{4\pi\varepsilon R_{ij}} \tag{3.5}$$

where e and ε are the elementary electron charge and dielectric constant of the media, respectively. R_{ij} is the distance between atoms i and j with charges q_i and q_j, respectively. If all atoms in the system are considered explicitly, the dielectric constant ε is set to be equal to that of the vacuum. In the simulations of solutions it is often possible to consider the solvent as a continuum. In this case, ε is set equal to the dielectric constant of the solvent, whereas only solute molecules and ions are explicitly considered. Such an approach enables the computational costs to be reduced and to increase the size of the system considered at the cost of neglecting molecular irregularities of the solvent. To allow flexibility, atomic charges can be assigned an effective partial charge, different from the formal valence.

The Lennard-Jones potential is commonly used to account for the short-range repulsion and dispersion interaction between atoms and molecules (U^{LJ}):

$$U^{LJ} = U^{\text{Repulsion}} + U^{\text{Dispersion}} = \sum_{i>j} \left(\left(\frac{A_{ij}}{R_{ij}} \right)^n - \left(\frac{B_{ij}}{R_{ij}} \right)^6 \right) \tag{3.6}$$

where R_{ij} is the interatomic distance; A_{ij}, B_{ij} and are parameters typically obtained by fitting experimental data; n is an integer typically chosen to be between 9 and 12. It should be noted, however, that the coefficients B_{ij} of the dispersion term can, in principle, be obtained by rigorous quantum mechanical calculations [14]. The above expression for short-range repulsion is very popular due to its simplicity. An exponential form provides a more accurate description of short range repulsion, and should be preferred in simulations of high pressure and high temperature systems:

$$U^{\text{Repulsion}} + U^{\text{Dispersion}} = \sum_{i>j} \left(C_{ij} \exp\left(-\frac{R_{ij}}{\rho_{ij}}\right) - \left(\frac{B_{ij}}{R_{ij}}\right)^6 \right) \tag{3.7}$$

where ρ_{ij}, C_{ij}, $B_{ij}$and are parameters typically obtained by fitting experimental data; The Morse potential is used to describe both intra- and inter-atomic interactions:

$$U^{\text{Morse}} = \sum_{i>j} D_{ij} \left(1 - \exp[-\beta_{ij}(R_{ij} - R_{\text{O}})]\right)^2 \tag{3.8}$$

where D_{ij} is the dissociation energy of the bond; β_{ij} is related to the vibrational frequency and R_{O} is the equilibrium bond distance.

Anions are surrounded by a loosely bound cloud of valence electrons. This electron density is readily deformed (polarized) upon changes in coordination shell, creating additional contributions to the bounding energy. The polarization can be modeled by allowing the displacement of the ions' negative charge (electron density) relative to the positivelt charged nuclei. The strength of the displacement is controlled by the harmonic potential. Since the displacement of the electron density depends on the coordination and at the same time exerts an additional induced force itself, the polarization has to be calculated self consistently by an iterative procedure [15]. Polarizable FFs significantly improve the accuracy of simulations. Polarization effects are particularly relevant for large anions with loosely bound outer electron shells and the simulations of material interfaces with different dielectric properties.

It is worth mentioning here a class of "reactive" FFs, which consider the possibility of bond dissociation [16]. For well-calibrated systems such a reactive force field could be a robust alternative to the expensive ab initio simulations. It has to be stressed that these parameters of these force fields are generally not "transferable". This means that the parameters have to be calibrated for a particular system and are not expected to give reliable results if applied to another system.

3.2.4.3 Quantum mechanical system description

In quantum mechanics, the state of electrons in the system is described by a wave function (WF, Ψ). The square of the wave function is equal to the electron density by definition $\rho = |\Psi|^2$. For a given nuclei position $\{R_I\}$, the ground state WF (Ψ_0) satisfies the stationary electronic Schrödinger equation:

$$\hat{H}\Psi_0(\{r_P\}, \{R_I\}) = E_0\Psi_0(\{r_P\}, \{R_I\}) \tag{3.9}$$

where $\hat{H}$ is the so called Hamiltonian operator which defines a second order differential equation; E_0 is the smallest eigenvalue of the Schrödinger equation, which has physical meaning of the ground state system energy; r_P and R_I are positions of electrons and nuclei, respectively. For real applications, the eq. (3.9) can only be solved in an approximate way. The most commonly used approaches are the Hartree-Fock [17] method and the density functional theory [18].

3.2.4.4 Hartree–Fock (HF) method

The HF method assumes that the total electronic WF of an N-electron system can be expressed as a determinant composed of N single electron WFs (the Slater determinant). A determinant has the property to change the sign upon exchange of arguments, which is the mathematical condition for the Pauli Exclusion Principle (so called exchange interaction). Upon substituting the Slater determinant into eq. (3.9), a system of N equations for non-interacting electrons is obtained. The electronic states are orthogonal and each electron feels the electrostatic field created by the nuclei and all the other electrons in the system. Since the potential acting on each electron depends on the electronic wave functions of the individual electrons, the system of equations has to be solved iteratively till the self-consistency is obtained. The HF method gives an exact description of the coulomb and exchange interactions. It ignores entirely the static and dynamic electrostatic correlations between electrons.

Calculations based on the HF method typically underestimate the bonding because the proper consideration of correlation effects would further lower the system energy. The so called post-HF methods, such as Møller–Plesset perturbation theory, configuration interaction theory, coupled cluster theory, etc. have been developed to account for electron correlation effects based on the HF reference solution. These methods can provide very accurate results, but the computational costs scale with the number of basis functions (see 2.4.8 Basis Sets) as M^6, and higher, making the use of these methods prohibitive for systems containing more than a few atoms [8].

3.2.4.5 Density functional theory (DFT)

The foundation of DFT is based on a strict mathematical proof that the ground state of a many-electron system is uniquely defined via a universal functional of the electron density [19]. The explicit mathematical expression for this functional is unknown, however. In principle, this functional should include terms for the kinetic energy of electrons, describe coulomb interactions, electron exchange, as well as static and dynamic electron-electron correlations. Only the coulomb interaction energy can be calculated explicitly from the electron density. Expressions for exchange and correlation energies, on the other hand, are known to a good approximation for a homogeneous electron gas. An explicit expression for the kinetic energy of electrons is defined for the electron's wave function but is not known for the electron density. To overcome these shortcomings, Kohn and Sham proposed to express the total electron density $\rho(r, \{R_I\})$ of the N-electron system, using N single-electron WFs $\psi_i^{KS}(r, \{R_I\})$, and to apply known expressions for the exchange-correlation energy of a homogeneous electron gas. This led to an expression for a single particle Kohn–Sham Hamiltonian $\hat{H}^{KS}$ [20]:

$$\rho(r) = |\Psi|^2 = \sum_{i=1}^{N} \left|\psi_i^{KS}(r, \{R_I\})\right|^2 \tag{3.10}$$

$$\hat{H}^{KS}\psi_i^{KS} = \varepsilon_i^{KS}\psi_i^{KS} \tag{3.11}$$

$$\hat{H}^{KS} = \underbrace{\overbrace{-\frac{\hbar^2}{2m_e}\nabla^2}^{\substack{\text{Kinetic Energy}\\ \text{of Kohn–Sham orbitals}}} + \overbrace{\overbrace{\hat{V}_{ext}(\{R_I\})}^{\text{Nuclei-Electron}} + \overbrace{\hat{V}_{Hartree}[\rho^{el}]}^{\text{Electron-Electron}}}^{\text{Coulomb Interaction}}}_{\text{Exact expression available}} + \underbrace{\overbrace{\hat{V}_{xc}[\rho^{el}]}^{\text{Quantum effects}}}_{\text{Approximation}} \tag{3.12}$$

where ψ_i^{KS} are the so called single particle Kohn–Sham orbitals, ε_i^{KS} are the eigenvalues, m_e is the mass of an electron. In this expression the electron-electron correlations and electron exchange are all lumped into the exchange-correlation term $\hat{V}_{xc}[\rho^{el}]$. The exact dependence of the *exchange* and *correlation* energy on the electron density ρ^{el} is not known, except for a homogeneous electron gas. In practical applications, $\hat{V}_{xc}[\rho^{el}]$ energies can be calculated within the so called local spin density approximation (LSDA) and its generalization (see below).

3.2.4.6 Exchange-correlation functionals and dispersion correction

Within the LSDA approximation, $\hat{V}_{xc}$ at each point in space is assumed to be equal to the energy of the homogeneous electron gas with the same (local) electron density. Despite its simplicity, the LSDA was very successful in describing the structural and electronic properties of solids and molecules. In general, the LSDA tends to underestimate atomic ionization energies and to overestimate binding energies. The performance of the LSDA functional can be further improved in several ways. First, a dependence on the density gradient (generalized gradient approximation, GGA) can be included, or even the higher order derivative of the electron density can be considered (meta-GGA). The latter functionals are sometime referred to as semi-local. Secondly, the exact exchange energy calculated using the Hartree-Fock approach and combined with exchange and correlation energy predicted by conventional GGA functionals in a semi-empirical way leads to the family of the so called "hybrid functionals". The GGA and hybrid-GGA functionals typically outperform LSDA for simple molecular systems [21, 22, 24, 25]. In solid state applications, neither LDA nor GGA is clearly preferable [26]. Both approaches have errors of comparable magnitude, though generally of opposite sign. LSDA is known to systematically underestimate lattice constants and coefficients of thermal expansion for solids, whereas GGA tends to overestimates these quantities. In contrast, LSDA tends to overestimate bulk moduli and phonon frequencies whereas GGA underestimates them [24].

A serious flaw of DFT in solid state applications is the underestimation of the band energy gap, which can be as high as 50 %. In contrast Hartree-Fock calculations tend to substantially overestimate the band gap. For these reasons, it is hoped that a hybrid-GGA functional could provide better predictions of band gaps in solids and thus give the more accurate description of the electronics states in the framework of DFT-approach. Another promising approach is the use of the so called orbital dependent functional [22]. A broad application of these methods is currently hindered by the high computational costs of the Hartree-Fock calculations performed with a plane wave basis set, which is the natural choice for periodic solids.

Density functional theory actually does not fully account for dispersion (van der Waals) forces between atoms. Such interactions can be very important for simulations of molecular sorption at mineral surfaces. An elegant and robust approach has been proposed by Grimme, in which the conventional DFT method is augmented by a semi-empirical dispersion correction (the so called DFT+D method) [27, 28]. The dispersion coefficients (e.g. last term in eq. (3.6), B_{ij}) consistent with the popular exchange correlation functionals have been derived for the entire periodic table based on quantum mechanical calculations [29].

Another challenge for DFT calculations, even with GGA and hybrid functionals, is the correct prediction of the absolute energy for electron orbitals of highly localized states such as *3d* states in transition metals or *4f* and *5f* states in lanthanides and actinides [23]. The simultaneous presence of multi-valence species (e.g. commonly the presence of Fe^{2+} and Fe^{3+} ions) adds extra complexity. The correct distribution (localization/de-localization) of electrons between ions of transition metals is very sensitive to the energy of the *3d* states. Even hybrid functionals often give unsatisfactory results. A substantial improvement in the accuracy of calculations at low computational overheads can be achieved based on the use of the Hubbard model [30] (so called DFT+U method). In this approach an extra empirical potential is imposed to obtain the correct energy of the localized electron states [31]. The parameters for the DFT+U method have to be calibrated based on experimental data, however.

Among a large number of different exchange-correlation functionals proposed during last decades, the PW91 [32], BLYP [33, 34] and PBE [35, 36] are the most popular and robust functionals of the GGA family used in solid state calculations. PBE0 [37] is probably the most promising hybrid functional increasingly used in condensed matter applications.

3.2.4.7 Pseudopotential approximation

At most thermodynamic conditions relevant to the Earth crust, the distribution of inner shell electrons in a solid is similar to that in an isolated atom. Only the valence electrons adjust their distribution to the changes in structural environment or chemical reactions. In the pseudopotential approach, the true electrostatic potential due to the nuclei core and inner electrons is replaced by an effective electrostatic potential [38]. The core electron orbitals are excluded from the simulations and the valence electrons are replaced by pseudo-WFs consistent with the pseudopotential. The pseudopotential and pseudo wave functions are constructed based on electronic structure calculation for an isolated atom in the ground and if necessary in excited states (See [39] for a recent review). The pseudo WFs are constructed to satisfy a number of conditions. They must have the same eigenvalues as true WFs, reproduce true WFs beyond a pre-defined cutoff distance from the nucleus and to be a smooth node-less function within this cutoff. The node-less form of pseudo-WF allows reducing the number of basis functions (see 2.4.8 Basis Set) that are required for an accurate representation of pseudo WFs and the electron density. The square of pseudo-WFs of the so-called norm conserving pseudopotentials integrates to the same electron density as true wave functions [40–42]. The last requirement is left out in the so called ultra-soft pseudopotentials. The ultra-soft pseudopotentials allow to reduce further the number of basis set functions by means of a more complicated computational algorithm [43].

3.2.4.8 Basis set

In both the DFT and HF approaches the WFs and electron density have to be expressed in a numerically convenient functional form (basis set). The plane wave basis set is a natural choice for simulations with periodic boundary conditions. This basis set is almost exclusively used in conjunction with the pseudopotentials, which allows high numerical accuracy to be achieved with a small number of plane wave basis functions. Exceptional numerical efficiency of a plane wave basis in DFT calculations is attained using the fast Fourier transform technique, which allows $n \cdot \log n$ scalability of the computational costs with the number of plane waves (n). Furthermore, plane wave basis sets converge in a smooth monotonic manner with the number of plane waves. This basis set is not associated with a particular atom (delocalized) and therefore provides an equally accurate representation of the WFs everywhere in the computational domain. To further improve computational efficiency, modern electronic structure codes combine reciprocal space representations of the electron density using plane waves with the real space Gaussian-like basis set for atomic orbitals. The later approaches allow the short range locality of electron distributions to be exploited [44].

Structural and electronic properties of a system are sensitive to the basis set in a different way. Good results for geometry can be obtained with a comparably small basis set. Energy and mechanical properties (elastic constants) are much more sensitive to the quality of the basis set. Thus, convergence of the system properties with respect to the basis set has to be tested systematically by increasing the number of basis functions until the required accuracy is obtained.

3.2.4.9 Ab initio molecular dynamics

According to the Ehrenfest theorem [45], an equation similar to eq. (3.3) also describes the dynamics of electrons. In principle, both systems of equations for collective motion of electrons and nuclei need to be solved simultaneously. It turns out that for many relevant phenomena, the motions of nuclei and electrons can be decoupled. The electrons are more than a factor of 10^3 lighter than the nuclei and thus move much faster. Solving eq. (3.3) for both nuclei and electrons simultaneously would need a very small integration step. Born-Oppenheimer (BO) and Car-Parrinello (CP) MD are two methods which deal with this issue.

In the BO approximation, electrons are assumed to be in their ground state for every instantaneous configuration of the nuclei. So, for every atomic configuration obtained by numerical integration of eq. (3.3), the ground state distribution of electrons and total energy is derived solving the corresponding Schrödinger equation (e.g. eq. (3.9) or eqs. (3.10)–(3.12). Knowing the total energy, the forces acting on the nuclei can be calculated and the dynamics of atoms can be predicted solving eq. (3.3). In CP-MD the propagation of the electronic WF close to the ground state is simulated by treating electrons as classical particles with a *fictitious* mass, The *fictitious* mass is chosen to be much larger than the actual mass of an electron but still smaller than the mass of the lightest nuclei in the system in order to maintain the adiabatic decoupling between the dynamics of nuclei and electrons [46]. In the CP approach, the Schrödinger equation is solved only once to obtain the ground state WFs at the beginning of the simulation. The time evolution of the WFs and nuclei is obtained by solving Newtons equations for electrons and nuclei simultaneously. Because of heavy fictitious electron mass, the equations of motion can be integrated with larger time step. The orthogonality of the WFs is maintained by mean of constraints using the method of Lagrange multipliers. Extensive benchmarking studies have shown that the value of the *fictitious* electron care must be carefully tested for each system. An inappropriate choice may lead to various numerical artifacts [47–50].

3.2.5 Thermodynamics of mineral surfaces

Surface structure can change with temperature and its composition depends on the chemical conditions. A general framework for predicting stable surface compositions and structure for arbitrary conditions based on ab initio simulations have been introduced by the pioneering work of Reuter and Scheffler [51]. The stability condition for a surface under given conditions is the minimum of the surface free energy $\gamma(T, P, N)$ subjected to eqs. (3.13)–(3.14):

$$\gamma(T, P, N) = \frac{1}{A}\left[G(T, P, N) - \sum_i N_i \mu_i(T, P)\right] \tag{3.13}$$

$$\mu_{\text{bulk}}(T, P) = \sum_j \tilde{N}_j \mu_j(T, P) \tag{3.14}$$

$G(T, P, N)$ is the Gibbs free energy of the system; $\mu_i(T, P)$ are chemical potential of the basis species; N_i and $\tilde{N}_j$ are their mole concentrations and A is the surface area. $\mu_{bulk}(T, P)$ is the chemical potential of the bulk. Eq. (3.14) imposes equilibrium of the surface with the bulk mineral.

The Gibbs free energy of the system is obtained as:

$$G(T, P, N) = E + F^{vib}(T, P, N) + PV(T, P, N) \tag{3.15}$$

The lattice energy E is the largest term in eq. (3.15). It is obtained in a straight-forward way by electronic structure calculations. For ambient conditions the $PV(T, P, N)$ term is small and can be entirely neglected. $F^{vib}(T, P, N)$ accounting for both the zero-point energy and thermal vibrations of atoms, can be calculated from the vibrational density of state $\sigma(V, \varpi)$:

$$F^{vib}(T, P, N) = \int d\varpi F^{vib}(T, V, \varpi)\sigma(V, \varpi) \tag{3.16}$$

$$F^{vib}(T, V, \varpi) = \hbar\varpi[1/2 + (e^{\beta\hbar\varpi} - 1)^{-1}] - \beta^{-1}[\beta\hbar\varpi(e^{\beta\hbar\varpi} - 1)^{-1} - \ln(1 - e^{-\beta\hbar\varpi})] \tag{3.17}$$

where $\beta = (k_B T)^{-1}$; $\hbar$ and k_B are the Plank and Boltzmann constants, respectively; ϖ and K are vibrational frequency and the thermodynamic temperature. The vibrational density of state is obtained either in the harmonic approximation [52] using finite differences (frozen phonon approximation) or based on perturbation density functional theory [53]. A full anharmonic vibrational density of states can be obtained by MD simulations [52]. Since $\sigma(V, \varpi)$ and $F^{vib}(T, V, \varpi)$ are obtained at constant volume, the volume has to match the pressure of interest.

Eqs. (3.13)–(3.17) provide a complete set of equations to determine the surface energy as a function of composition and chemical potential (eqs. (3.13)–(3.14)), as well as temperature and pressure (eqs. (3.15)–(3.17)). In many applications the influence of thermal vibrations can be neglected. In this case eq. (3.13) reduces to:

$$\gamma(T, P, N) = \frac{1}{A}[E(N) - \sum_i N_i\mu_i(T, P)]. \tag{3.18}$$

3.2.6 Acid-base properties of surfaces

3.2.6.1 Intrinsic acidity of the sorption sites

In natural environments, mineral surfaces participate in acid-base equilibria with water rich geochemical fluids. The interpretation of experimental titration data is often based on empirical models for the acidity of surface sites ($\equiv ss$). It is currently not possible to derive the acidity of individual surface groups at the mineral surfaces experimentally. Recently, a numerical approach has been proposed which allows the prediction of protonation constants for surface sites using DFT and taking into account explicit solvent and temperature effects by means of ab initio MD simulations [54–56].

In this method, free energy $\Delta_{dp}A$ of a proton transfer reaction (eq. (3.19)) is derived by the thermodynamic integration of the ensemble averaged vertical protonation gap $\langle \Delta_{dp}E \rangle_{H(\lambda)}$:

$$\equiv \mathrm{ssH} + \mathrm{H_2O_{aq}} \leftrightarrow \quad \equiv \mathrm{ss^-} + \mathrm{H_3O^+_{aq}} \tag{3.19}$$

$$\Delta_{dp}A = \int_0^1 d\lambda \, \langle \Delta_{dp}E \rangle_{H(\lambda)}. \tag{3.20}$$

The quantity $\Delta_{dp}E$ is the potential energy of the proton transfer reaction for a given configuration. The simulations were performed in an ensemble defined by a λ-dependent Hamiltonian:

$$H_\lambda = (1 - \lambda)H_0 + \lambda H_1, \tag{3.21}$$

where H_0 and H_1 are the Hamiltonians describing the system of adduct and product of the proton transfer reaction in eq. (3.19).

The equilibrium constant (pK_a) of the de-protonation reaction (eq. (3.19)) is calculated in the low temperature limit neglecting the difference in the zero point vibration energy of the proton in $\equiv ss$H and $\mathrm{H_3O^+_{aq}}$ complexes as:

$$2.3\, k_B T \mathrm{p}K_a = k_B T Ln \frac{a_{\equiv ss^-} a_{H_3O^+}}{a_{H_2O} a_{\equiv ssH}} = \Delta_{dp}A + k_B T \ln[c^0 \Lambda^3_{H^+}] \tag{3.22}$$

where k_B is the Boltzmann constant, T is temperature, a indicates the activity, c^0 = 1 mol L^{-1} is the standard state concentration and Λ_{H^+} is the thermal wavelength of a proton. The last term in eq. (3.22) is a thermochemical correction constant equal to −3.2 pK units. This term accounts for the configurational entropy of the proton in water [54].

In practical calculations, the integral in eq. (3.20) is approximated by numerical integration (e.g. Simpson's formula). Three and more integration points are necessary for the evaluation of the integral. This implies that the simulations have to be performed for system of adducts ($\lambda \equiv 0$), products ($\lambda \equiv 1$), as well as several intermediate states described by the mixed Hamiltonian ($0 < \lambda < 1$). These mixed-Hamiltonian systems do not have chemical equivalents but rather provide a thermodynamic path between well-defined thermodynamic states. The simulations are straight forward if the acidity of the surface site is lower than the acidity of the water. If the water is more acidic than the surface, a spontaneous protonation of the surface site may occur during the simulation, especially for H_1. A practical solution to this issue is to apply restrains to the interatomic distances in the H_2O molecules. Since the restrains are applied for both adduct and product states, the energy due to the restraints cancels out.

3.2.6.2 Effect of the electric double layer

The method described above can predict the acid-base properties of surface sites at the pH of zero charge in equilibrium with pure water. Some mineral phases may not be stable under such conditions, or only exist in equilibrium with a mineralized fluid. Surface charge created due to protonation-de-protonation reactions at a surface in contact with electrolyte solutions induce electrostatic potentials, which in turn modify the spatial distribution of anions and cations near the interface and lead to the formation of a so called electrical double layer (EDL). Direct simulations of electrolyte solutions at mineral surfaces from first principles are not feasible due to the long relaxation times involved. Usually, empirical surface complexation models [57] or course-grain simulation approaches are applied to predict the surface charge and the ion distribution at the interface [58]. Surface charge σ_0 and potential Ψ are related via capacitance C:

$$C = \frac{d\sigma_0}{d\Psi} \tag{3.23}$$

In empirical sorption models the capacitance is considered as a fitting parameter. Recently, an approach has been proposed allowing to estimate the capacitances of the EDL at the charged mineral surfaces based on ab initio MD simulations and to compare these results with an empirical model fitted to the experimental data [59]. One of these approaches relies on the energy of the conduction band minimum as function of the surface charge (U_{CBM}), calculated a follow:

$$eU_{CBM} = -\Delta A_{+H^{+}+e^{-}} - \mu_{H^{+}}^{g,\circ} - \Delta E_{ZPE} \tag{3.24}$$

where $\Delta A_{+H^{+}+e^{-}}$ is the free energy for a simultaneous reversible insertion of an electron in the solid and a proton in the fluid phase. This quantity is estimated by an ab initio thermodynamic integration approach similar to the one used for the calculation of pK (eqs. (3.20)–(3.23)). $\mu_{H^{+}}^{g,\circ}$ and ΔE_{ZPE} are standard chemical potentials of a proton in the gas phase and an energy correction for the zero point energy of the proton in water, both of which are system independent constants [59]. The capacitance is calculated as:

$$C^{-1} = \frac{dU_{CBM}}{d\sigma_0}. \tag{3.25}$$

3.2.7 Simulation packages

Several open source and commercial simulation packages for the solid state applications are available. ABINIT (www.abinit.org), QuantumExpresso (www.quantum-espresso.org) and CPMD (www.cpmd.org) are plane wave based DFT codes which perform basis electronic structure calculations (total energy, forces stress tensor) and provide access to other material properties based on perturbation theory. Compared to CPMD, the ABINIT has the more efficient implementation of k-point sampling. QuantumExpresso offers advanced functionality for the modelling of phonon and NMR spectra. CP2K (www.cp2k.org) is a hybrid code using both plane-wave and Gaussian basis sets. Its particular strength is the simulation of large disordered systems. VASP (www.vasp.at), CASTEP (www.castep.org) and WIEN2K (wien2k.org) are commercial simulation packages available for academic use at relatively low cost. WIEN2K-implements one of the most accurate methods for band structure calculations and is often used to benchmark the results obtained with pseudopotential calculations.

3.3 Case studies

3.3.1 Diamond

Diamond is the hardest mineral so far known. Each carbon atom in the diamond structure is in the sp^3 hybridization state, which allows the formation of perfect tetrahedral coordination. The hardness of diamond is orientation dependent and is highest in the [111] direction [60]. The (001) surface of diamond is the slowest growing surface in the chemical vapor deposition process used for manufacturing synthetic diamonds and thus is of high industrial importance [61]. The surface properties, including hardness, strongly depend on the surface treatment. The most studied surfaces of diamond are (001) and (111) [62–68]. The clean (001) surface is known to form a $(001) - 2 \times 1$ reconstruction (Fig. 3.3). Neighboring atoms at the surface come together to form double bonded dimers ($C = C$) with a bond distance 1.37 Å. For comparison, the $C - C$ bond in the bulk diamond structure is about 1.5 Å. The dimers are arranged in rows along [110] directions. The $C = C$ surface dimers contribute to bonding and antibonding states (π and π^*) with a gap of about 1.3 eV, making the surface semiconducting. For comparison, the bulk diamond is an insulator with a band gap of 5.5 eV [66]. The filled surface states are placed energetically within the valence band of the bulk diamond. The surface relaxation results in a shortening of the interlayer distances and a buckling of the near surface atomic planes [62].

The monohydrogenated $(001) - 2 \times 1 - 2H$ surface of diamond is believed to be the most stable diamond surface under ambient conditions. In hydrated surfaces, the π bounding in the surface dimer is replaced by a σ_{s-p} OH bond. The bond length of the $C - C$ surface dimers increase to about 1.6 Å, reflecting the change in the bonding

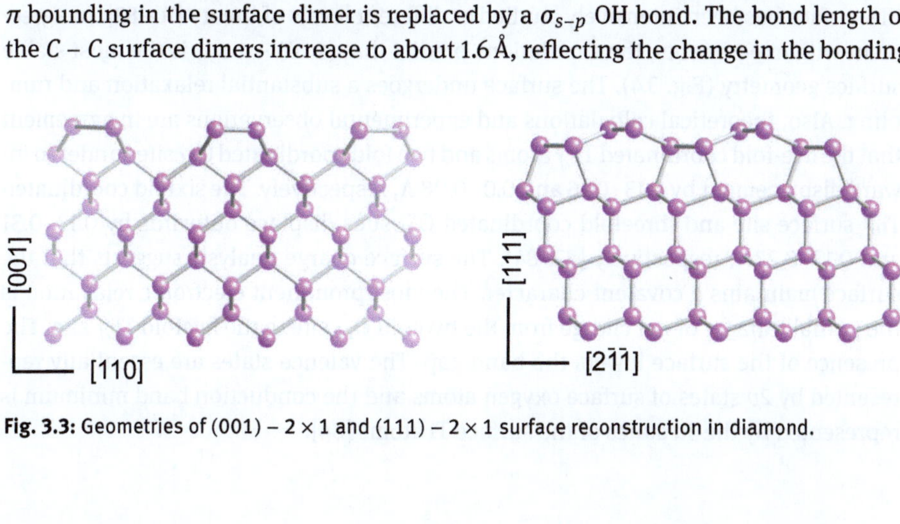

Fig. 3.3: Geometries of $(001) - 2 \times 1$ and $(111) - 2 \times 1$ surface reconstruction in diamond.

mechanism. The surface hydrogenation counter balances the surface potential and results in much smaller relaxation and buckling of the near surface atomic planes. The band gap on the $(001) - 2 \times 1 - 2H$ surface is lowered due to an upward shift of the valence states by about 2 eV. Similar structural changes are observed for the hydroxylation of (001) surface. The energy of the hydroxylated surface reconstruction $(001) - 2 \times 1$ OH lies in between the surface energy of the hydrogenated and the clean $(001) - 2 \times 1$ surface [64].

The clean unreconstructed (111) surface of diamond is unstable and undergoes the so called Pandey-chain reconstruction $(111) - 2 \times 1$. This type of reconstruction was first described by Pandey for the $Si(111) - 2 \times 1$ surface [69]. The surface atoms in the Pandey chain form 1.43 Å long $C - C$ bonds. Surface π states are located energetically near the Fermi level of the bulk diamond. Due to the weak interaction between the Pandey chains, the splitting of bonding and antibonding surface states is weak and the surface is expected to have a semimetallic character. Hydrogenation of the (111) surface stabilizes an unreconstructed $(111) - 1 \times 1 - H$ surface termination [70, 71]. The electronic properties of the $(111) - 1 \times 1 - H$ surface are very similar to the those of the $(001) - 2 \times 1 - 2H$ one. The hydrogenation removes the surface state from the valence gap and the surface becomes semiconducting.

3.3.2 Rutile

One of the most intensively studied surfaces of oxides is that of rutile (TiO_2). The rutile structure is a model for several industrially relevant metal oxides. It is also a preferred material for experimental studies because high quality surfaces can be prepared by sputtering and annealing. Thus surfaces of rutile have been characterized by most of the available experimental techniques. Excellent compilations of experimental and modelling results can be found in recent reviews [3, 72–75]. The (110) is the most stable and also the least reactive stoichiometric surface of rutile. The structure of the (110) rutile surface has been intensively studied at different levels of theory [76–83]. In agreement with experiments, the simulations predict that $(110) - 1 \times 1$ is the most stable surface geometry (Fig. 3.4). The surface undergoes a substantial relaxation and rumpling. Also, theoretical calculations and experimental observations are in agreement that the five-fold coordinated Ti_{5f} atoms and two fold coordinated O_{2f} sites undergo inward displacement by 0.13–0.16 and 0.0–0.08 Å, respectively. The sixfold coordinated Ti_{6f} surface site and threefold coordinated O_{3f} sites displace outwards by 0.13–0.31 and 0.13–0.22 Å, respectively [82, 84]. The surface charge analysis suggests that the surface maintains a covalent character. The most prominent electronic relaxation is the partial transfer of net charge from the twofold O_{2f} site to the fivefold Ti_{5f} site. The presence of the surface lowers the band gap. The valence states are essentially represented by *2p* states of surface oxygen atoms and the conduction band minimum is represented by the *3d* states of the surface Ti atoms [84].

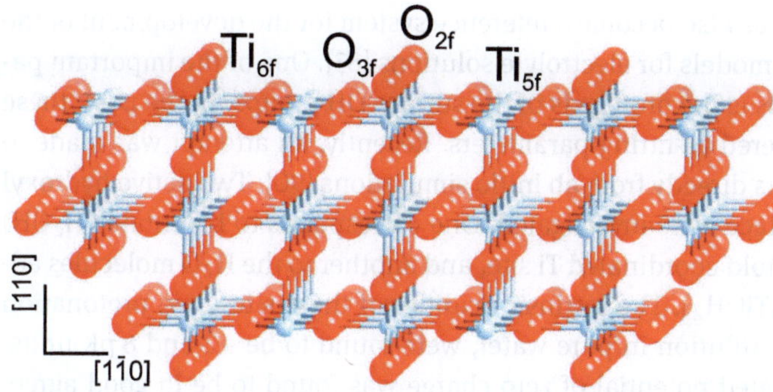

Fig. 3.4: (110) – 1 × 1 surface of rutile. Ti and O atoms are shown by blue and red spheres, respectively.

An accurate description of the *3d* electronic states of Ti atoms is challenging for standard DFT methods. Comparative studies between DFT calculations with LDA, GGA, hybrid exchange-correlation functionals and DFT+U approach indicate that accurate predictions of surface energies, and in particular the electronic structure of the surfaces with defects, using conventional LDA or GGA functionals can give unsatisfactory results [79, 85, 86].

The growth of the (110) rutile surface has been modelled by combining ab initio total energy calculations, MD simulations based on coordination dependent interaction potentials and on-the-fly kinetic Monte Carlo simulations [81, 87]. The ab initio simulations were used to obtain activation energies for the different atom deposition pathways and to calibrate the parameters for the force field for the MD simulations. The activation energies were used to calculate deposition rates probabilities applied in KMC simulations. The system specific force field parameters were applied to simulate of the surface growth directly from atomistic MD simulations. These simulations are computationally time consuming and are used as a benchmark for the more efficient KMC approach.

The interaction of water with the (110) surface has been a topic of continuous research by both theoretical and experimental methods. One of the central issues in these studies is whether water adsorbs as a molecule or by a dissociative mechanism. Most experimental studies suggest that water adsorption on the ideal (110) surface takes place by a non-dissociative molecular mechanism [88]. The theoretical studies are inconclusive [72, 76, 77, 80, 89]. The simulations suggest that the relative stability of the phys-sorbed and chem-sorbed water at the TiO_2 surface strongly depends on the thickness of the TiO_2 slab used for the simulations, the size of the supercell as well as the symmetry of the upper and lower slab surface [80]. Furthermore, it was suggested that the stability of the chemisorbed water can be further increased due to hydrogen bonding. Thus, multilayer water adsorption should favor the water dissociation.

The rutile surface has also become a reference system for the development of the sorption complexation models for electrolyte solutions [83]. One of the important parameters in such models is the protolysis of the surface OH groups. In the past, these quantities were considered as fitting parameters. Recently, an attempt was made to evaluate these constants directly from ab initio simulations [90]. Two active hydroxyl groups at the (110) surface were investigated. One corresponds to bidentate (Ti_2OH) site attached to the six fold coordinated Ti site, and another to the H_2O molecules adsorbed at the Ti_{5f} site ($TiOH_2$). The intrinsic acidities of these sites, e.g. protonation constants at an infinite dilution in pure water, were found to be −2 and 8 pK units, respectively. The estimated potential of zero charge was found to be in good agreement with the experimental measurements. A further step taken towards a description of electrolyte solutions at mineral interfaces was the evaluation of the EDL capacitance for a NaF electrolyte solution. The simulations show that the capacities vary with the protonation state of the surface. The capacitance of the negatively charged (de-protonated) surface was found to be lower than that of the positively charged (protonated) surface. This can be explained by the fact that fluorine interacts weaker with its solvation shell than Na due to a larger ionic radius. Therefore, the effective dielectric constant of media with a F^- rich double layer is larger than the one of a Na^+ rich double layer. Accordingly, the capacitance of the positively charged surface was predicted to be larger than that of negatively charged one.

Other low index surfaces of rutile are more reactive and show larger structural variability [82]. Different $(011) - 2 \times 1$ surface reconstructions have been suggested based on simulations in vacuum [91–94]. The observed differences in the structure of the surface could be related to the differences in the simulation setup and the methods used. The simulations suggest that water adsorbs on $(011) - 2 \times 1$ rutile surface by a dissociation mechanism [89, 95]. It was also argued by [89] that chemisorption of water on the $(011) - 2 \times 1$surface can lead to stabilization of $(011) - 1 \times 1$ geometry.

3.3.3 Pyrite

Pyrite is one of the most abundant sulfide minerals in geological formations. Oxidation of pyrite in mining tails is responsible for the formation of acidic waste water. The oxidation process may follow different reaction mechanisms depending on the environmental conditions. The presence of sulfide and iron oxidizing bacteria can substantially increase the rates of oxidative pyrite dissolution. Clay minerals and aqueous cations Al^{3+}, Fe^{3+}, Cu^{2+} may also act as catalysts. Thus the stability and reactivity of pyrite surfaces has been the subject of intensive research [96–115]. Simulations suggest that the clean (100) pyrite surface maintains 1×1 geometry (Fig. 3.5) [98, 109, 111, 114]. The surface is formed by the breaking of Fe-S bonds, whereby the S-S dimers remain intact. The Fe atoms in the top most layer are fivefold coordinated. The surface relaxation is manifested by a shortening of the Fe-S bonds oriented normal

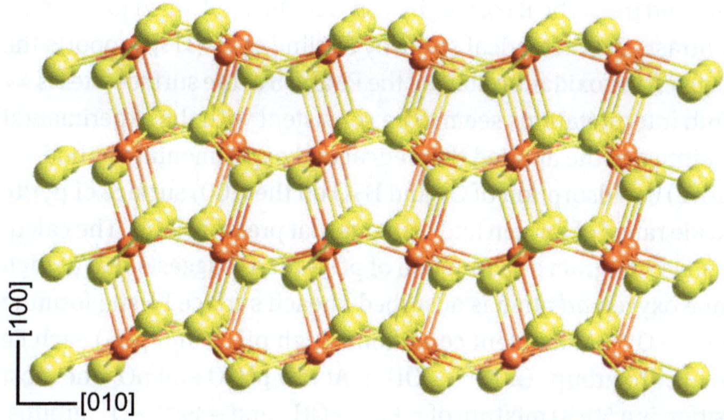

Fig. 3.5: (100) – 1 × 1 surface of pyrite. Fe and S atoms are shown by red and yellow spheres, respectively.

to the surface plane. In contrast, the changes in the S-S bonds and in-plane Fe-S bonds are minor. The obtained electronic structure suggests that the surface is non-metallic and diamagnetic.

H_2O molecules were found to adsorb on top of the five coordinated Fe sites by a non-dissociative mechanism. The orientation of H_2O molecules at low coverage is controlled by weak -H..S- hydrogen bonds. As the water coverage increases, neo formed -H..OH- hydrogen bonds start to dominate the geometry of the adsorbed water layer [111]. Recent studies of multilayer adsorption suggest that the presence of bulk water leads to the shortening of Fe-O_w bonds. It can be explained by a stronger polarization of the near surface H_2O molecules due to multilayer adsorption and thereby to a stronger Fe-O_w interaction [99].

Pyrite shows imperfect cleavage parallel to (100), suggesting that this surface should incorporate a significant number of defects. These defects should play a key role in surface oxidation mechanism. Monomeric S sites, either dissociated S_2 dimers or S atoms, adsorbed at fivefold coordinated surface Fe sites, are considered to be the most important active sites for the oxidation of pyrite surfaces. These defect couples can be formed by the dissociation of S-S dimers followed by the adsorption of released S atoms at Fe sites by two different mechanisms:

$$S_2^{2-} \rightarrow S^- + S^- \tag{3.26}$$

$$Fe_{surf}^{2+} + S^- \rightarrow Fe_{surf}^{3+} - S^{2-} \tag{3.27}$$

or

$$S_2^{2-} \rightarrow S^{2-} + S^0 \tag{3.28}$$

$$Fe_{surf}^{2+} + S^0 \rightarrow Fe_{surf}^{4+} - S^{2-} \tag{3.29}$$

Combined experimental and theoretical studies by von Oertzen [115] are in favor of the first mechanism. In contrast, the theoretical study by Stirling at al. [113] supports the second mechanism in which the oxidation state of the Fe and S at the surface sites is 4+ and 2-, respectively. Both interpretations seem to be consistent with the experimental data within the uncertainties of the applied theoretical and experimental methods.

Reactive (dissociative) co-adsorption of O_2 and H_2O on the (100) surface of pyrite has been studied at a wide range of oxygen and water partial pressure [110]. The calculated surface stability phase diagram as a function of pO_2–pH_2O suggests that at high pO_2 (and low pH_2O) one oxygen *ad* atom is adsorbed at each surface Fe site forming ferryl-oxo species $\equiv Fe^{4+} - O^{2-}$. At ambient conditions (high pH_2O and pO_2) each Fe surface site forms ferric-hydro groups ($\equiv Fe^{3+} - OH^-$). At low pH_2O and pO_2 the most stable surface composition is a 50:50 mixture of $\equiv Fe^{3+} - OH^-$ and $\equiv Fe^{4+} - O^{2-}$ groups.

Full surface coverage with $\equiv Fe^{3+} - OH^-$ groups (favorable at ambient conditions) was considered as the starting point for sulfur oxidation at the (100) surface of pyrite. It was suggested that surface S-S dimers oxidize to SO_4^{2-} through a chain of intermediate H_2O and O_2 adsorption steps by a dissociative mechanism. Each reaction step was found to be exothermic with activation barriers not exceeding 25 kcal/mol, suggesting that the process is slow but still feasible at room temperature.

Periodic quantum mechanical studies of other pyrite surfaces are scarce [102, 103]. In agreement with experimental observations, the (110), (111) and (210) pyrite surfaces have higher surface energies than the (100) face. The (110), (111) and (210) surfaces have by far more variable surface structures (e.g. 4, 5 and 6 coordinated Fe sites) and thus higher reactivity.

3.3.4 Magnetite

Magnetite is the strongest magnetic mineral found on the Earth so far. It is an abundant accessory mineral in igneous, metamorphic and sedimentary rocks. The ferrimagnetic property of magnetite is used for paleomagnetic reconstructions. Its oxidation to hematite buffers oxygen fugacity in geochemical systems. Under ambient conditions and higher temperatures, Fe_3O_4 adopts an inverse spinel structure built on a cubic close packed sub-lattice of O^{2-} atoms. One eighth of the available tetrahedral sites is occupied by Fe^{3+} cations and one half of octahedral sites is occupied by equal amounts of Fe^{2+} and Fe^{3+}. However, this formal assignment of an integer charge (2+ and 3+) to the Fe atoms at octahedral sites is misleading. In fact, the "extra" electron, assigned to Fe^{2+} belongs to the conduction band of minor spin states and therefore it is delocalized over all octahedral sites. In this context it is more appropriate to assign a fractional charge (+2.5) to the Fe at octahedral positions. The charge ordering (localization of electrons at octahedral Fe sites in such a way that non-equivalent Fe^{2+} and Fe^{3+} sites emerge) indeed takes place at 121 K (the so called Verwey transition) [116–118]. Below this temperature a cubic spin-polarized magnetite (half metallic phase)

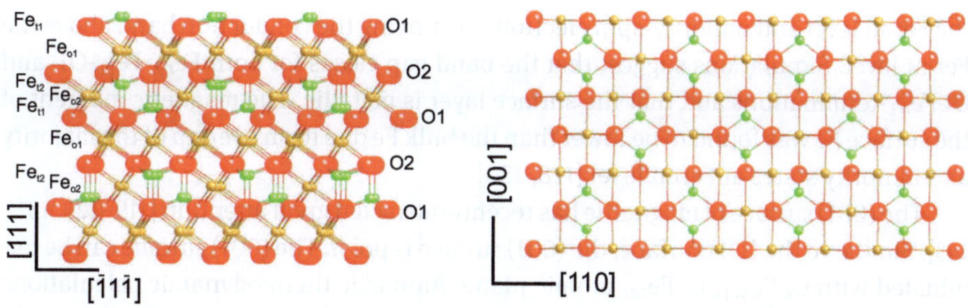

Fig. 3.6: (111) and (001) surfaces of magnetite. Octahedral and tetrahedral iron is shown by brown and green spheres, respectively. Oxygen atoms are red.

transforms into a semiconducting monoclinic phase with structurally non-equivalent octahedral Fe^{2+} and Fe^{3+} sites. The symmetry of the low-temperature Fe_3O_4 phase has certain similarities with the structural reconstruction observed at the (001) magnetite surface, which is considered by some authors as a surface analog of the Verwey transition [119–122]. The (111) and (001) are dominant facets of magnetite crystals resulting in a broad family of cube-octahedral forms. The stability of the different morphologies has been predicted as the function of oxygen fugacity and redox conditions [123].

The stacking sequence of atomic planes in magnetite along the [111] direction is shown in Fig. 3.6. Any unreconstructed (111) termination of magnetite has a non-zero dipole moment, and thus, according the electrostatic arguments of Tasker [124], can only be stabilized by substantial structural re-arrangement of the surface layers or co-adsorption. The stability, geometry and composition of the (111) magnetite surface for different terminations have been investigated in detail [122, 125–131]. The calculations suggest that Fe_{o2}-Fe_{t1}-O1- termination dominates in a wide range of oxygen fugacities, whereas the O1-Fe_{o1}- terminated surface is only stable in a narrow region of high partial oxygen fugacity close to atmospheric conditions. Both terminations undergo strong inward relaxation, which helps to partially recover the coordination of the surface atoms and reduce the surface dipole moment. It has been confirmed by different modelling approaches that the relaxation of a Fe_{o2}-Fe_{t1}-O1- terminated surface leads to a change in the ordering of -Fe_{t1}-O1- atomic planes near the surface [125, 130]. Adsorbed H_2O preferably binds to Fe_{O2}- sites. Dissociation of H_2O is an energetically favored exothermic process with a moderate activation energy barrier [129]. It is worth mentioning that Fe_{o2}-Fe_{t1}-O1- and Fe_{t1}-O1- terminations were predicted to have surface energies close to one another and could coexist on disordered surfaces [125, 126, 128]. The energy of an Fe_{t1}-O1- termination was found to be lowered further by dissociative hydrogen adsorption [125, 127].

Structural rearrangements at the magnetite surface (111) have profound consequences for the electronic and magnetic properties of the surface layer. Bulk magnetite is a half-metal, with majority-spin electrons occupying the valence band below the Fermi level and minority-spin electrons occupying the conduction band above the Fermi level. Simulations suggest that the band gap closes for both Fe_{o2}-Fe_{t1}-O1- and O1-Fe_{o1}-terminations and that the surface layer is metallic. The magnetic moment of the surface Fe was found to be lower than the bulk Fe due to an overlap of the majority and minority states at Fermi level [128, 130].

The (001) surface of magnetite has recently attracted great attention [119–123, 131–143]. Similar to the (111) surface, the (001) surface is polar. The (001) surface can be terminated with O, Fe_{oct}, or Fe_{tet} atomic plane. Atomistic thermodynamic calculations suggest that the Fe_{oct} terminated surface is dominant, whereas the Fe_{tet} terminated surface could potentially be stable in a narrow range of low oxygen fugacity. It is widely accepted that the (001) surface of magnetite undergoes $(001) - \sqrt{2} \times \sqrt{2} - R45°$ reconstruction. The origin and underlying changes of electronic and magnetic properties of the surface layer are still a matter of debate [119, 120, 122, 128, 133, 140–142, 144].

Studies by [133, 138, 141, 142, 144] suggest that upon $(001) - \sqrt{2} \times \sqrt{2} - R45°$ reconstruction the surface layer of magnetite undergoes a half-metal to metal phase transition. The nature of this relaxation is essentially explained by a Jahn-Teller distortion of the surface Fe atoms. A study by [140] suggests that the (001) surface retains half-metal character similar to the bulk magnetite. In contrast, recent studies argue that $(001) - \sqrt{2} \times \sqrt{2} - R45°$ reconstruction is a result of charge ordering in the near surface layer, structurally similar to a low temperature Verwey phase transition [119, 121, 122]. These studies predict a semiconducting state of the surface layer. Another study confirms the formation of a semiconducting surface, but explains the origin of $(001) - \sqrt{2} \times \sqrt{2} - R45°$ reconstruction by presence of Fe vacancies rather than charge ordering [120]. None of the reported experimental and theoretical studies provide an ultimate proof for the nature of the (001) magnetite surface. A recent comparative simulation of bulk magnetite with various exchange-correlation functionals suggests that the predicted electronic structure depends strongly on the functional used and can lead to qualitatively different results [145]. This could be one of the reasons for diverging interpretations of the simulation results.

The electron conductivity of the (001) surface layer of magnetite was estimated based on a model of small polaron hopping between Fe^{3+} and Fe^{2+} sites in the octahedral sub-lattice. The calculations predicted a polaron hoping rate of up to two orders of magnitude lower than that in the bulk. The reduced polaron mobility was explained by a high energy barrier for Fe^{3+}–Fe^{2+} reorganization in the surface layer. It was suggested that the availability of Fe^{2+} at the surface and the reduced conductivity of the surface layer might be the rate limiting for redox reaction involving magnetite [143].

3.3.5 Phyllosilicates

Phyllosilicate minerals are omnipresent in soils and other sedimentary rocks. Platelets of phyllosilicate minerals are composed of alternating sheets of cations having an octahedral configuration (O-sheets) and sheets of Si-tetrahedra (T-sheets). The most common sequences formed are TOT and TO layers, also called 2:1 and 1:1 phyllosilicates. Heterovalent substitutions in octahedral and tetrahedral sheets result in a net negative charge of the TOT-layers. This permanent charge is compensated by hydrated and/or bare cation alkali and alkali-earth elements. A high surface area and permanent surface charge are responsible for outstanding sorption capacities of phyllosilicate minerals. Two surface types, namely the basal plane and the edges, can be distinguished. The basal plane, formed by oxygen in the tetrahedral sheets (in 2:1 phyllosilicates) or by OH groups (in 1:1 phyllosilicates) at the octahedral sheets. The basal plane is known to be chemically inert in a wide range of environmental conditions. The sorption properties of basal surfaces have been studied by classical simulations with empirical FFs over several decades. (e.g. recent reviews [5, 146, 147]). Ab initio simulations are scarcer [148–157]. Edge-surfaces readily participate in acid-base reactions and are responsible for the pH-buffering capacity of phyllosilicates. Pyrophyllite and kaolinite are two substitution-free 2:1 and 1:1 clay minerals, respectively. They can be considered as structural prototypes for a wide range of phyllosilicates. Cleaved edge surfaces of phyllosilicate adsorb water by dissociation mechanism [158]. Ab initio simulations suggest that (010) and (110) are the most stable hydrated edge facets of phyllosilicates [158–163]. The surface structure and protonation schema of (010), $(0\bar{1}0)$, (110), $(\bar{1}\bar{1}0)$ of cis-vacant pyrophyllite considered as prototype of 2:1 phyllosilicates is shown in Fig. 3.7. Acid-base properties of the OH groups at (110) and (010) edge sites of pyrophyllite and kaolinite have been directly estimated based on an ab initio thermodynamic integration approach [164–166]. These data provide a fundamental structural basis for the formulation of empirical surface complexation models [57]. Due to the potential use of clays as a buffer material in the disposal of conventional and radioactive waste, sorption mechanisms of actinides, fission products and heavy metals need to be understood. Atomistic simulations provide insights into the mechanisms of these processes at the molecular level and help in the interpretation of spectroscopic results [139, 167, 168]. The simulations suggest that depending on the thermodynamic conditions divalent transition metals can form bidentate complexes at the edge surface or can be incorporated into the octahedral sheet of phyllosilicates [170, 171]. The predicted geometries of the surface complexes agree well with the structural parameters derived from the spectroscopic studies.

3.4 Outlook

Quantum mechanical modelling of mineral surfaces is a rapidly evolving field. It is a powerful tool which provides in-depth structural and chemical information concerning surface processes occurring at an atomistic scale. Most of the applications in the past have been limited to rather simple binary systems. However, even these "simple" systems show very complex behavior attributable to atomic relaxations in the surface

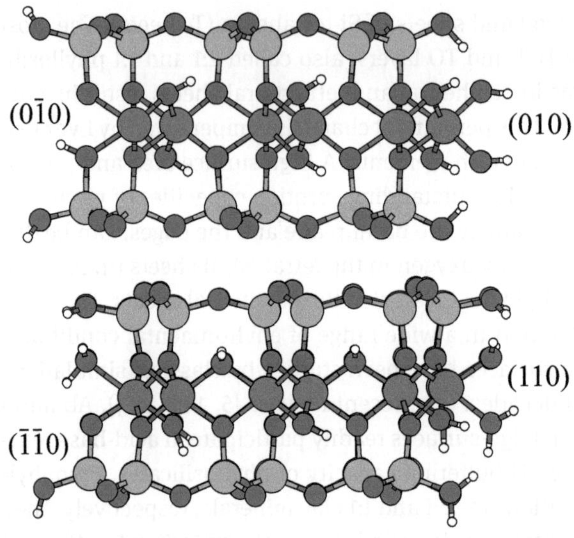

Fig. 3.7: Surface structure and protonation schema of (010), (0$\bar{1}$0), (110), ($\bar{1}\bar{1}$0) of cis-vacant pyrophyllite considered as prototype of 2:1 phyllosilicates. Si atoms are brown, Al atoms are green, O atoms are red, H atoms are white.

layer. The further development of theory, numerical algorithms and increasing computing performance will inevitably broaden the scope of application. Some challenging and intriguing applications would be ab initio simulations of disordered systems, mineral-fluid interfaces, surface reactivity and atomistic thermodynamic modelling from first principles. Simulations of these systems would substantially improve the interpretation of spectroscopic data and model development. Nowadays state-of-the-art ab initio simulations can handle systems with as many as several 10^4 atoms [172]. This opens the door to ab initio simulations of "reality sized" nanomaterials for which surface phenomena dominate the macroscopic behavior.

Bibliography

[1] Oura K, Lifshits VG, Saranin A, Zotov AV, Katayama M. Surface Science: An Introduction: Springer, 2003.
[2] Noguera C. Physics and Chemistry at Oxide Surfaces: Cambridge University Press, 1996.
[3] Noguera C. Polar oxide surfaces. Phys Condens Matter 2000, 12, R367–R410.
[4] Diebold U, Li S-C, Schmid M. Oxide Surface Science. In: Leone SR, Cremer PS, Groves JT, Johnson MA, Richmond G, editors. Annu Rev Phys Chem 2010, 61, 129–148.
[5] Geysermans P, Noguera C. Advances in atomistic simulations of mineral surfaces. J Mater Chem 2009, 19, 7807–7821.
[6] Wood EA. Vocabulary of Surface Crystallography. J Appl Phys 1964, 35, 1306–1312.

[7] Marx D, Hutter J. Ab Initio Molecular Dynamics: Basic Theory and Advanced Methods. Cambridge University Press, 2009.
[8] Foresman JB, Frisch A. Exploring Chemistry with electronic structure methods: a guide to using gaussian. Gaussian Inc., 1996.
[9] Allen MP, Tildesley DJ. Computer simulations of liquids. Oxford: Clarendon Press, 1987.
[10] Frenkel D, Smit B. Understanding Molecular Simulation. Academic Press, 2002.
[11] Martin RM. Electronic structure calculations. Basic theory and practical methods. Cambridge University Press, 2004.
[12] Blochl PE. Electrostatic decoupling of periodic images of plane-wave-expanded densities and derived atomic point charges. J Chem Phys 1995, 103, 7422–7428.
[13] Martyna GJ, Tuckerman ME. A reciprocal space based method for treating long range interactions in ab initio and force-field-based calculations in clusters. J Chem Phys 1999, 110, 2810–2821.
[14] Casimir HBG, Polder D. The influence of retardation on the London-van der Waals forces. Phys Rev 1948, 73, 360–372.
[15] Rick SW, Stuart SJ. Potentials and algorithms for incorporating polarizability in computer simulations. In: Lipkowitz KB, Boyd DB, editors. Reviews in Computational Chemistry 2002, 18, 89–146.
[16] Duin ACTv, Strachan A, Stewman S, Zhang Q, Xu X, Goddard WA. ReaxFFSiO Reactive Force Field for Silicon and Silicon Oxide Systems. J Phys Chem A 2003, 107, 3803–3811.
[17] Jensen F. Introduction to Computational Chemistry. 2nd ed, Wiley, 2006.
[18] Parr RG, Yang W. Density-functional theory of atoms and molecules. New York, Oxford: Oxford University Press, 1989.
[19] Hohenberg P, Kohn W. Inhomogeneous electron gas. Phys Rev 1964, 136, 864–871.
[20] Kohn W, Sham LJ. Self-consistent equation including exchange and correlation effects. Phys Rev 1965, 140, A1133–A8.
[21] Sousa SF, Fernandes PA, Ramos MJ. General performance of density functionals. J Phys Chem A 2007, 111, 10439–10452.
[22] Kummel S, Kronik L. Orbital-dependent density functionals: Theory and applications. Rev Mod Phys 2008, 80, 3–60.
[23] Furche F, Perdew JP. The performance of semilocal and hybrid density functionals in 3d transition-metal chemistry. J Chem Phys 2006, 124, 044103.
[24] Csonka GI, Perdew JP, Ruzsinszky A, Philipsen PHT, Lebegue S, Paier J, et al. Assessing the performance of recent density functionals for bulk solids. Phys Rev B 2009, 79, 155107.
[25] Cramer CJ, Truhlar DG. Density functional theory for transition metals and transition metal chemistry. Phys Chem Chem Phys 2009, 11, 10757–10816.
[26] Kurth S, Perdew JP, Blaha P. Molecular and solid-state tests of density functional approximations: LSD, GGAs, and meta-GGAs. Int J Quantum Chem 1999, 75, 889–909.
[27] Grimme S. Semiempirical GGA-Type Density Functional Constructed with a Long-Range Dispersion Correction. J Comput Chem 2007, 27, 1787–1798.
[28] Grimme S. Accurate description of van der Waals complexes by density functional theory including empirical corrections. J Comput Chem 2004, 25, 1463–1473.
[29] Grimme S, Antony J, Ehrlich S, Krieg H. A consistent and accurate ab initio parametrization of density functional dispersion correction (DFT-D) for the 94 elements H-Pu. J Chem Phys 2010, 132, 154104.
[30] Hubbard J. Electron Correlations in Narrow Energy Bands. Proc R Soc Lond A Math Phys Sci 1963, 285, 238–257.

[31] Anisimov VI, Aryasetiawan F, Lichtenstein AI. First-principles calculations of the electronic structure and spectra of strongly correlated systems: The LDA+U method. J Phys Condens Matter 1997, 9, 767–808.
[32] Perdew JP, Wang Y. Accurate and simple analytic representation of the electron-gas correlation energy. Phys Rev B 1992, 45, 13244–13249.
[33] Becke AD. Density-functional exchange-energy approximation with correct asymptotic behavior. Phys Rev A 1988, 38, 3098–3100.
[34] Lee C, Yang W, Parr RG. Development of the Colle-Salvetti correlation-energy formula into a functional of the electron density. Phys Rev B 1988, 37, 785–789.
[35] Perdew JP, Burke K, Ernzerhof M. Errata: Generalized gradient approximation made simple [Phys Rev Lett 1996, 77, 3865]. Phys Rev Lett 1997, 78, 1396.
[36] Perdew JP, Burke K, Ernzerhof M. Generalized gradient approximation made simple. Phys Rev Lett 1996, 77, 3865–3868.
[37] Adamo C, Barone V. Toward reliable density functional methods without adjustable parameters: The PBE0 model. J Chem Phys 1999, 110, 6158
[38] Pickett WE. Pseudopotential methods in condensed matter applications. Comput Phys Rep 1989, 9, 115–198.
[39] Dolg M, Cao X. Relativistic Pseudopotentials: Their Development and Scope of Applications. Chem Rev 2012, 112, 403–480.
[40] Goedecker S, Teter M, Hutter J. Separable dual-space Gaussian pseudopotentials. Phys Rev B 1996, 54, 1703–1710.
[41] Troullier N, Martins JL. Efficient pseudopotentials for plane-wave calculations. Phys Rev B 1991, 43, 1993–2006.
[42] Bachelet GB, Hamann DR, Schlüter M. Pseudopotentials that work: From H to Pu. Phys Rev B 1982, 26, 4199–4228.
[43] Vanderbilt D. Soft self-consistent pseudopotentials in a generalized eigenvalue formalism. Phys Rev B 1990, 41, 7892–7895.
[44] Goedecker S. Linear scaling electronic structure methods. Rev Mod Phys 1999, 71, 1085–1123.
[45] Ehrenfest P. Bemerkung über die angenäherte Gültigkeit der klassischen Mechanik innerhalb der Quantenmechanik. Z Phys 1927, 45, 455–457.
[46] Car R, Parrinello M. Unified approach for molecular dynamics and density-functional theory. Phys Rev Lett 1985, 55, 2471–2474.
[47] Grossman JC, Schwegler E, Draeger EW, Gygi F, Galli G. Toward assessment of the accuracy of density functional theory for first principle simulations of water. J Chem Phys 2004, 120, 300–311.
[48] Schwegler E, Grossman JC, Gygi F, Galli G. Towards an assessment of the accuracy of density functional theory for first principles simulations of water. II. J Chem Phys 2004, 121, 5400–5409.
[49] Kuo I-FW, Mundy CJ, McGrath MJ, Siepmann JI, VandeVondele J, Sik M, et al. Liquid water from first principles: investigation of different sampling approaches. J Phys Chem B 2004, 108, 12990–12998.
[50] Kuo I-FW, Mundy CJ, McGrath MJ, Siepmann JI. Time-dependent properties of liquid water: A comparison of Car-Parrinello and Born-Oppenheimer molecular dynamics simulations. J Chem Theory Comput 2006, 2, 1274–1281.
[51] Reuter K, Scheffler M. Composition, structure, and stability of RuO_2 (110) as a function of oxygen pressure. Phys Rev B 2001, 65, 035406.
[52] Dove MT. Introduction to Lattice Dynamics. Cambridge University Press, 1993.

[53] Giannozzi P, Gironcoli Sd, Pavone P, Baroni S. Ab initio calculation of phonon dispersions in semiconductors. Phys Rev B 1991, 43, 7231–7242.
[54] Costanzo F, Sulpizi M, Valle RGD, Sprik M. The oxidation of tyrosine and tryptophan studied by a molecular dynamics normal hydrogen electrode. J Chem Phys 2011, 134, 244508.
[55] Cheng J, Sulpizi M, Sprik M. Redox potentials and pKa for benzoquinone from density functional theory based molecular dynamics. J Chem Phys 2009, 131, 154504.
[56] Sulpizi M, Sprik M. Acidity constants from vertical energy gaps: density functional theory based molecular dynamics implementation. Phys Chem Chem Phys 2008, 10, 5238–5249.
[57] Tournassat C, Grangeon S, Leroy P, Giffaut AE. Modeling specific pH dependent sorption of divalent metals on montmorillonite surfaces. A review of pitfalls, recent achievements and current challenges. Am J Sci 2013, 313, 395–451.
[58] Churakov SV, Labbez C, Pegado L, Sulpizi M. Intrinsic acidity of surface sites in calcium-silicate-hydrates and its implication to their electrokinetic properties. J Phys Chem C 2014, 118, 11752–11762.
[59] Cheng J, Sprik M. The electric double layer at a rutile TiO_2 water interface modelled using density functional theory based molecular dynamics simulation. J Phys Condens Matter 2014, 26, 244108.
[60] Blank V, Popov M, Pivovarov G, Lvova N, Gogolinsky K, Reshetov V. Ultrahard and superhard phases of fullerite C-60: comparison with diamond on hardness and wear. Diamond and Related Materials 1998, 7, 427–431.
[61] Lee ST, Lin ZD, Jiang X. CVD diamond films: nucleation and growth. Mater Sci Eng R-Rep 1999, 25, 123–154.
[62] Alfonso DR, Drabold DA, Ulloa SE. Structural, electronic, and vibrational properties of diamond(100), diamond(111), and diamond(110) surfaces from ab-initio calculations. Phys Rev B 1995, 51, 14669–14685.
[63] Bechstedt F, Stekolnikov AA, Furthmuller J, Kackell P. Origin of the different reconstructions of diamond, Si, and Ge(111) surfaces. Phys Rev Lett 2001, 87, 016103.
[64] Sque SJ, Jones R, Briddon PR. Structure, electronics, and interaction of hydrogen and oxygen on diamond surfaces. Phys Rev B 2006, 73, 085313.
[65] De La Pierre M, Bruno M, Manfredotti C, Nestola F, Prencipe M, Manfredotti C. The (100), (111) and (110) surfaces of diamond: an ab initio B3LYP study. Mol Phys 2014, 112, 1030–1039.
[66] Ristein J. Surface science of diamond: Familiar and amazing. Surface Science 2006, 600, 3677–3689.
[67] Walter S, Bernhardt J, Starke U, Heinz K, Maier F, Ristein J, et al. Geometry of the (2x1) reconstruction of diamond(111). J Phys Condens Matter 2002, 14, 3085–3092.
[68] Kruger P, Pollmann J. Dimer reconstruction of diamond, Si, and Ge (001) surfaces. Phys Rev Lett 1995, 74, 1155–1158.
[69] Pandey KC. New π-Bonded Chain Model for Si(111)-(2×1) Surface. Phys Rev Lett 1981, 47, 1913–1917.
[70] Pate BB, Hecht MH, Binns C, Lindau I, Spicer WE. Photoemission and photon-stimulated ion desorption studies of diamond(111)-hydrogen. J Vac Sci Technol 1982, 21, 364–367.
[71] Pate BB, Waclawski BJ, Stefan PM, Binns C, Ohta T, Hecht MH, et al. The diamond (111) surface – a dilemma resolved. Physica B & C 1983, 117, 783–785.
[72] Pang CL, Lindsay R, Thornton G. Structure of Clean and Adsorbate-Covered Single-Crystal Rutile TiO_2 Surfaces. Chem Rev 2013, 113, 3887–3948.
[73] Sun C, Liu L-M, Selloni A, Lu GQ, Smith SC. Titania-water interactions: a review of theoretical studies. J Mater Chem 2010, 20, 10319–10334.
[74] Diebold U. Structure and properties of TiO_2 surfaces: a brief review. Appl Phys A 2003, 76, 681–687.

[75] Pang CL, Lindsay R, Thornton G. Chemical reactions on rutile TiO_2(110). Chem Soc Rev 2008, 37, 2328–2353.
[76] Bandura AV, Sykes DG, Shapovalov V, Troung TN, Kubicki JD, Evarestov RA. Adsorption of water on the TiO_2 (rutile) (110) surface: A comparison of periodic and embedded cluster calculations. J Phys Chem B 2004, 108, 7844–7853.
[77] Harris LA, Quong AA. Molecular chemisorption as the theoretically preferred pathway for water adsorption on ideal rutile TiO_2(110). Phys Rev Lett 2004, 93, 086105.
[78] Lindan PJD, Harrison NM, Gillan MJ, White JA. First-principles spin-polarized calculations on the reduced and reconstructed TiO_2 (110) surface. Phy Rev B 1997, 55, 15919–15927.
[79] Morgan BJ, Watson GW. A DFT+U description of oxygen vacancies at the TiO_2 rutile (110) surface. Surface Science 2007, 601, 5034–5041.
[80] Perron H, Vandenborre J, Domain C, Drot R, Roques J, Simoni E, et al. Combined investigation of water sorption on TiO_2 rutile (110) single crystal face: XPS vs. periodic DFT. Surface Science 2007, 601, 518–527.
[81] Vernon L, Kenny SD, Smith R, Sanville E. Growth mechanisms for TiO_2 at its rutile (110) surface. Phys Rev B 2011, 83, 075412.
[82] Ramamoorthy M, Vanderbilt D, Kingsmith RD. 1st-principles calculations of the energetics of stoichiometric TiO_2 surfaces. Phys Rev B 1994, 49, 16721–16727.
[83] Zhang Z, Fenter P, Cheng L, Sturchio NC, Bedzyk MJ, Predota M, et al. Ion adsorption at the rutile-water interface: Linking molecular and macroscopic properties. Langm 2004, 20, 4954–4969.
[84] Labat F, Baranek P, Adamo C. Structural and electronic properties of selected rutile and anatase TiO_2 surfaces: An ab initio investigation. J Chem Theory Comput 2008, 4, 341–352.
[85] Tilocca A, Selloni A. DFT-GGA and DFT+U Simulations of Thin Water Layers on Reduced TiO_2 Anatase. J Phys Chem C 2012, 116, 9114–9121.
[86] Yuan F, Lu S, Xu W, Zhang H, Ning T. LSDA plus U calculations of the electronic and optical properties of rutile TiO_2(110) vs (011)-2 × 1 surfaces. Comput Mater Sci 2014, 90, 1–6.
[87] Sanville EJ, Vernon LJ, Kenny SD, Smith R, Moghaddam Y, Browne C, et al. Surface and interstitial transition barriers in rutile (110) surface growth. Phys Rev B 2009, 80, 235308.
[88] Diebold U. The surface science of titanium dioxide. Surface Science Rep 2003, 48, 53–229.
[89] Aschauer U, Selloni A. Structure of the Rutile TiO_2(011) Surface in an Aqueous Environment. Phys Rev Lett 2011, 106, 166102.
[90] Cheng J, Sprik M. Acidity of the Aqueous Rutile TiO_2(110) Surface from Density Functional Theory Based Molecular Dynamics. J Chem Theory Comput 2010, 6, 880–889.
[91] Beck TJ, Klust A, Batzill M, Diebold U, Di Valentin C, Selloni A. Surface structure of TiO_2(011)-(2x1). Phy Rev Lett 2004, 93, 036104.
[92] Beck TJ, Klust A, Batzill M, Diebold U, Di Valentin C, Tilocca A, et al. Mixed dissociated/molecular monolayer of water on the TiO_2(011)-(2 × 1) surface. Surface Science 2005, 591, L267–L272.
[93] Kubo T, Orita H, Nozoye H. Surface structures of rutile TiO_2 (011). J Am Chem Soc 2007, 129, 10474–10478.
[94] Yuan F, Lu S, Xu W, Zhang H. First-principles study of structural, electronic, and optical properties of the rutile TiO_2(011)-2 × 1 surface. Surface Science 2014, 621, 88–93.
[95] Yuan F, Zhang H, Lu S, Xu W. The hydroxylated and reduced rutile TiO_2(011)-2 × 1 surfaces: A first-principles study. Surface Science 2014, 628, 126–131.
[96] Alfonso DR. Computational Investigation of FeS_2 Surfaces and Prediction of Effects of Sulfur Environment on Stabilities. J Phys Chem C 2010, 114, 8971–8980.
[97] Blanchard M, Wright K, Gale JD, Catlow CRA. Adsorption of $As(OH)_3$ on the (001) surface of FeS_2 pyrite: A quantum-mechanical DFT study. J Phys Chem C 2007, 111, 11390–11396.

[98] Cai J, Philpott MR. Electronic structure of bulk and (001) surface layers of pyrite FeS_2. Comp Mater Sci 2004, 30, 358–363.

[99] Chen J, Long X, Chen Y. Comparison of Multilayer Water Adsorption on the Hydrophobic Galena (PbS) and Hydrophilic Pyrite (FeS_2) Surfaces: A DFT Study. J Phys Chem C 2014, 118, 11657–11665.

[100] Corkhill CL, Warren MC, Vaughan DJ. Investigation of the electronic and geometric structures of the (110) surfaces of arsenopyrite (FeAsS) and enargite (Cu_3AsS_4). Miner Mag 2011, 75, 45–63.

[101] de Lima GF, de Oliveira C, de Abreu HA, Duarte HA. Water Adsorption on the Reconstructed (001) Chalcopyrite Surfaces. J Phys Chem C 2011, 115, 10709–10717.

[102] Hung A, Muscat J, Yarovsky I, Russo SP. Density-functional theory studies of pyrite FeS_2 (111) and (210) surfaces. Surface Science 2002, 520, 111–119.

[103] Hung A, Muscat J, Yarovsky I, Russo SP. Density-functional theory studies of pyrite FeS_2(100) and (110) surfaces. Surface Science 2002, 513, 511–524.

[104] Murphy R, Strongin DR. Surface reactivity of pyrite and related sulfides. Surface Science Rep 2009, 64, 1–45.

[105] Nair NN, Schreiner E, Marx D. Glycine at the pyrite-water interface: The role of surface defects. J Am Chem Soc 2006, 128, 13815–13826.

[106] Rosso KM, Becker U, Hochella MF. The interaction of pyrite {100}surfaces with O_2 and H_2O: Fundamental oxidation mechanisms. Am Mineral 1999, 84, 1549–1561.

[107] Rosso KM, Vaughan DJ. Reactivity of sulfide mineral surfaces. Sulfide Mineralolgy and Geochemistry 2006, 61, 557–607.

[108] Rosso KM, Vaughan DJ. Sulfide mineral surfaces. Sulfide Mineralolgy and Geochemistry 2006, 61, 505–556.

[109] Sacchi M, Galbraith MCE, Jenkins SJ. The interaction of iron pyrite with oxygen, nitrogen and nitrogen oxides: a first-principles study. Phys Chem Chem Phys 2012, 14, 3627–3633.

[110] Sit PHL, Cohen MH, Selloni A. Interaction of Oxygen and Water with the (100) Surface of Pyrite: Mechanism of Sulfur Oxidation. J Phys Chem Lett 2012, 3, 2409–2414.

[111] Stirling A, Bernasconi M, Parrinello M. Ab initio simulation of water interaction with the (100) surface of pyrite. J Chem Phys 2003, 118, 8917–8926.

[112] Stirling A, Bernasconi M, Parrinello M. Ab initio simulation of H_2S adsorption on the (100) surface of pyrite. J Chem Phys 2003, 119, 4934–4939.

[113] Stirling A, Bernasconi M, Parrinello M. Defective pyrite (100) surface: An ab initio study. Phys Rev B 2007, 75, 165406.

[114] Von Oertzen GU, Harmer SL, Skinner WM. XPS and ab initio calculation of surface states of sulfide minerals: pyrite, chalcopyrite and molybdenite. Mol Simul 2006, 32, 1207–1212.

[115] von Oertzen GU, Skinner WM, Nesbitt HW. Ab initio and x-ray photoemission spectroscopy study of the bulk and surface electronic structure of pyrite (100) with implications for reactivity. Phys Rev B 2005, 72, 235427.

[116] Leonov I, Yaresko AN, Antonov VN, Korotin MA, Anisimov VI. Charge and orbital order in Fe_3O_4. Phys Rev Lett 2004, 93, 146404.

[117] Zhang Z, Satpathy S. Electron-states, magnetism, and the verwey transition in magnetite. Phys Rev B 1991, 44, 13319–13331.

[118] Senn MS, Wright JP, Attfield JP. Charge order and three-site distortions in the Verwey structure of magnetite. Nature 2012, 481, 173–176.

[119] Bernal-Villamil I, Gallego S. Charge order at magnetite Fe_3O_4(001): surface and Verwey phase transitions. J Phys Condens Matter 2015, 27, 012001.

[120] Bliem R, McDermott E, Ferstl P, Setvin M, Gamba O, Pavelec J, et al. Subsurface cation vacancy stabilization of the magnetite (001) surface. Science 2014, 346, 1215–1218.

[121] de la Figuera J, Novotny Z, Setvin M, Liu T, Mao Z, Chen G, et al. Real-space imaging of the Verwey transition at the (100) surface of magnetite. Phys Rev B 2013, 88, 161410.
[122] Lodziana Z. Surface verwey transition in magnetite. Phys Rev Lett 2007, 99, 206402.
[123] Santos-Carballal D, Roldan A, Grau-Crespo R, de Leeuw NH. A DFT study of the structures, stabilities and redox behaviour of the major surfaces of magnetite Fe_3O_4. Phys Chem Chem Phys 2014, 16, 21082–21097.
[124] Tasker PW. The stability of ionic crystal surfaces. J Phy Condens Matter 1979, 12, 4977–4984.
[125] Ahdjoudj J, Martinsky C, Minot C, Van Hove MA, Somorjai GA. Theoretical study of the termination of the Fe_3O_4 (111) surface. Surface Science 1999, 443, 133–153.
[126] Paul M, Sing M, Claessen R, Schrupp D, Brabers VAM. Thermodynamic stability and atomic and electronic structure of reduced Fe_3O_4 (111) single-crystal surfaces. Phys Rev B 2007, 76, 075412.
[127] Yang T, Wen XD, Huo CF, Li YW, Wang J, Jiao H. Structure and energetics of hydrogen adsorption on Fe_3O_4(111). J Mol Catal A Chem 2009, 302, 129–136.
[128] Yu XH, Huo CF, Li YW, Wang JG, Jiao HJ. Fe_3O_4 surface electronic structures and stability from GGA plus U. Surface Science 2012, 606, 872–879.
[129] Zhou CG, Zhang QF, Chen L, Han B, Ni G, Wu JP, et al. Density Functional Theory Study of Water Dissociative Chemisorption on the Fe_3O_4(111) Surface. J Phys Chem C 2010, 114, 21405–21410.
[130] Zhu L, Yao KL, Liu ZL. First-principles study of the polar (111) surface of Fe_3O_4. Phys Rev B 2006, 74, 035409.
[131] Pabisiak T, Kiejna A. Fe adsorption on hematite (alpha-Fe_2O_3) (0001) and magnetite (Fe_3O_4) (111) surfaces. J Chem Phys 2014, 141, 134707.
[132] Bartelt NC, Nie S, Starodub E, Bernal-Villamil I, Gallego S, Vergara L, et al. Order-disorder phase transition on the (100) surface of magnetite. Phys Rev B 2013, 88, 235436.
[133] Fonin M, Dedkov YS, Pentcheva R, Ruediger U, Guentherodt G. Magnetite: a search for the half-metallic state. J Phys Condens Matter 2007, 19, 315217.
[134] Kendelewicz T, Kaya S, Newberg JT, Bluhm H, Mulakaluri N, Moritz W, et al. X-ray Photoemission and Density Functional Theory Study of the Interaction of Water Vapor with the Fe_3O_4(001) Surface at Near-Ambient Conditions. J Phys Chem C 2013, 117, 2719–2733.
[135] Mulakaluri N, Pentcheva R. Hydrogen Adsorption and Site-Selective Reduction of the Fe_3O_4(001) Surface: Insights From First Principles. J Phys Chem C 2012, 116, 16447–16453.
[136] Mulakaluri N, Pentcheva R, Scheffler M. Coverage-Dependent Adsorption Mode of Water on Fe_3O_4(001): Insights from First Principles Calculations. J Phys Chem C 2010, 114, 11148–11156.
[137] Mulakaluri N, Pentcheva R, Wieland M, Moritz W, Scheffler M. Partial Dissociation of Water on Fe_3O_4(001): Adsorbate Induced Charge and Orbital Order. Phys Rev Lett 2009, 103, 176102.
[138] Parkinson GS, Mulakaluri N, Losovyj Y, Jacobson P, Pentcheva R, Diebold U. Semiconductor-half metal transition at the Fe_3O_4(001) surface upon hydrogen adsorption. Phys Rev B 2010, 82, 125413.
[139] Sun X, Kurahashi M, Pratt A, Yamauchi Y. First-principles study of atomic hydrogen adsorption on Fe_3O_4(100). Surface Science 2011, 605, 1067–1073.
[140] Cheng C. Structure and magnetic properties of the Fe_3O_4(001) surface: Ab initio studies. Phys Rev B 2005, 71, 052401.
[141] Pentcheva R, Moritz W, Rundgren J, Frank S, Schrupp D, Scheffler M. A combined DFT/LEED-approach for complex oxide surface structure determination: Fe_3O_4(001). Surface Science 2008, 602, 1299–1305.
[142] Pentcheva R, Wendler F, Meyerheim HL, Moritz W, Jedrecy N, Scheffler M. Jahn-Teller stabilization of a "polar" metal oxide surface: Fe_3O_4(001). Phys Rev Lett 2005, 94, 126101.

[143] Skomurski FN, Kerisit S, Rosso KM. Structure, charge distribution, and electron hopping dynamics in magnetite (Fe_3O_4) (100) surfaces from first principles. Geochim Cosmochim Acta 2010, 74, 4234–4248.

[144] Fonin M, Pentcheva R, Dedkov YS, Sperlich M, Vyalikh DV, Scheffler M, et al. Surface electronic structure of the Fe_3O_4(100): Evidence of a half-metal to metal transition. Phys Rev B 2005, 72, 104436.

[145] Noh J, Osman OI, Aziz SG, Winget P, Bredas J-L. A density functional theory investigation of the electronic structure and spin moments of magnetite. Sci Technol Adv Mater 2014, 15, 044202.

[146] Ignacio Sainz-Diaz C. Applications of computational atomistic methods to phyllosilicates. Layered Mineral Structures and Their Application in Advanced Technologies. EMU Notes in Mineralogy 2011, 11, 203–236.

[147] Suter JL, Anderson RL, Greenwell HC, Coveney PV. Recent advances in large-scale atomistic and coarse-grained molecular dynamics simulation of clay minerals. J Mater Chem 2009, 19, 2482–2493.

[148] Churakov SV, Kosakowski G. An ab initio molecular dynamics study of hydronium complexation in Na-montmorillonite. Philos Mag 2010, 90, 2459–2474.

[149] Boek ES, Sprik M. Ab initio molecular dynamics study of the hydration of a sodium smectite clay. J Phys Chem B 2003, 107, 3251–3256.

[150] Chatterjee A, Ebina T, Onodera Y, Mizukami F. Effect of exchangeable cation on the swelling property of 2 : 1 dioctahedral smectite – A periodic first principle study. J Chem Phys 2004, 120, 3414–3424.

[151] Liu X, Meijer EJ, Lu X, Wang R. Ab initio molecular dynamics study of fe-containing smectites. Clays Clay Miner 2010, 58, 89–96.

[152] Clausen P, Andreoni W, Curioni A, Hughes E, Plummer CJG. Water Adsorption at a Sodium Smectite Clay Surface: an Ab Initio Study of the First Stage. J Phys Chem C 2009, 113, 15218–15225.

[153] Clausen P, Andreoni W, Curioni A, Hughes E, Plummer CJG. Adsorption of Low-Molecular-Weight Molecules on a Dry Clay Surface: An Ab Initio Study. J Phys Chem C 2009, 113, 12293–12300.

[154] Demontis P, Masia M, Suffritti GB. Water Nanoconfined in Clays: The Structure of Na Vermiculite Revisited by Ab Initio Simulations. J Phys Chem C 2013, 117, 15583–15592.

[155] Demontis P, Masia M, Suffritti GB. Peculiar Structure of Water in Slightly Superhydrated Vermiculite Clay Studied by Car-Parrinello Molecular Dynamics Simulations. J Phys Chem C 2014, 118, 7923–7931.

[156] Stackhouse S, Coveney PV, Sandre E. Plane-wave density functional theoretic study of formation of clay-polymer nanocomposite materials by self-catalyzed in situ intercalative polymerization. J Am Chem Soc 2001, 123, 11764–11774.

[157] Tunega D, Benco L, Haberhauer G, Gerzabek MH, Lischka H. Ab initio molecular dynamics study of adsorption sites on the (001) surfaces of 1 : 1 dioctahedral clay minerals. J Phys Chem B 2002, 106, 11515–11525.

[158] Churakov SV. Ab initio study of sorption on pyrophyllite: Structure and acidity of the edge sites. J Phys Chem B 2006, 110, 4135–4146.

[159] Churakov SV. Structure and dynamics of the water films confined between edges of pyrophyllite: A first principle study. Geochim Cosmochim Acta 2007, 71, 1130–1144.

[160] Bickmore BR, Rosso KM, Nagy KL, Cygan RT, Tadanier CJ. Ab initio determination of edge surface structures for dioctahedral 2:1 phyllosilicates: implications for acid-base for reactivity. Clays Clay Miner 2003, 51, 359–371.

[161] Liu X, Lu X, Meijer EJ, Wang R, Zhou H. Atomic-scale structures of interfaces between phyllosilicate edges and water. Geochim Cosmochim Acta 2012, 81, 56–68.
[162] Liu X, Lu X, Wang R, Meijer EJ, Zhou H, He H. Atomic scale structures of interfaces between kaolinite edges and water. Geochim Cosmochim Acta 2012, 92, 233–242.
[163] Martins DMS, Molinari M, Goncalves MA, Mirao JP, Parker SC. Toward Modeling Clay Mineral Nanoparticles: The Edge Surfaces of Pyrophyllite and Their Interaction with Water. J Phys Chem C 2014, 118, 27308–27317.
[164] Liu X, Cheng J, Sprik M, Lu X, Wang R. Surface acidity of 2:1-type dioctahedral clay minerals from first principles molecular dynamics simulations. Geochim Cosmochim Acta 2014, 140, 410–417.
[165] Liu X, Lu X, Sprik M, Cheng J, Meijer EJ, Wang R. Acidity of edge surface sites of montmorillonite and kaolinite. Geochim Cosmochim Acta 2013, 117, 180–190.
[166] Tazi S, Rotenberg B, Salanne M, Sprik M, Sulpizi M. Absolute acidity of clay edge sites from ab-initio simulations. Geochim Cosmochim Acta 2012, 94, 1–11.
[167] Liu X, Meijer EJ, Lu X, Wang R. Ab initio molecular dynamics study of Fe-Containing smectites. Clays Clay Miner 2010, 58, 89–96.
[168] Liu X, Meijer EJ, Lu X, Wang R. Ab initio molecular dynamics study of Fe-containing smectites. Clays Clay Miner 2010, 58, 89–96.
[169] Liu X, Lu X, Meijer EJ, Wang R, Zhou H. Acid dissociation mechanisms of $Si(OH)_4$ and $Al(H_2O)_6^{3+}$ in aqueous solution. Geochim Cosmochim Acta 2010, 74, 510–416.
[170] Churakov SV, Daehn R. Zinc Adsorption on Clays Inferred from Atomistic Simulations and EXAFS Spectroscopy. Environ Sci Technol 2012, 46, 5713–5719.
[171] Alexandrov V, Rosso KM. Insights into the Mechanism of Fe(II) Adsorption and Oxidation at Fe-Clay Mineral Surfaces from First-Principles Calculations. J Phys Chem C 2013, 117, 22880–22886.
[172] Khaliullin RZ, VandeVondele J, Hutter J. Efficient Linear-Scaling Density Functional Theory for Molecular Systems. J Chem Theory Comput 2013, 9, 4421–4427.

Luca Bindi and Paul J. Steinhardt

4 Natural quasicrystals: A new frontier in mineralogy and its impact on our understanding of matter and the origin of the solar system

4.1 Prelude to quasicrystals

With the discovery of the first natural quasicrystal in 2009, a new frontier in mineralogy has opened that could lead to new discoveries in geoscience, astronomy, condensed matter physics, and materials engineering. For the first time, minerals have been discovered that violate the symmetry restrictions of conventional crystallography. That nature could accomplish this without human intervention was unexpected, requiring the existence of petrological processes never considered previously. The fact that the quasicrystals were found in a meteorite formed in the earliest moments of the solar system means these processes have been active for over 4.5 billion years and influenced the mineral composition of the first objects to condense around the Sun. Finding quasicrystals formed in these extreme environments informs the longstanding debate about the stability and robustness of quasicrystals among condensed matter physicists. Finally, the discovery inspires further searches for quasicrystals and other new forms of matter not seen in the laboratory previously, which may provide valuable new materials for physics and engineering.

The discovery of natural quasicrystals is not the first time that mineralogy has had a broad impact in science. The pioneering work of Rene-Just Haüy establishing that minerals can have only limited facets, dihedral angles and symmetries provided some of the first evidence that matter is composed of irreducible units (which Haüy called molecules), providing arguably the first evidence for the atomic hypothesis. The same analysis applied to human-made materials as well and, hence, profoundly influenced solid state physics for the next two centuries. Haüy's work also motivated the first use of group theory in science, which set the foundations of crystallography and deeply affected many realms of chemistry and physics. Later, the discovery of X-ray diffraction in 1912 by von Laue and its first application by him and by William and Lawrence Bragg (1914) to analyse simple crystals brought Haüy's concepts into a modern quantum context.

The key principle that governed Haüy's ideas and crystallography for the next two centuries was lattice periodicity. Periodicity was considered to be an essential feature of all solids with non-random atomic arrangements. From that principle followed rigorous mathematical theorems that restrict rotational symmetry, rules that were considered to be evident and unbreakable.

At the beginning of the 20th century, however, mineralogists reported striking observations challenging the validity of the well-known law of rational indices [1–3] derived from assuming periodicity. By careful high-precision goniometry measurements of several crystal facets of the mineral calaverite ($Au_{1-x}Ag_xTe_2$, with $0 \leq x \leq 0.33$), these authors discovered the possibility of *incommensurability* in crystals. Only after nearly half a century did the morphology of calaverite begin to be understood [4, 5], based on invoking four indices rather than the three that suffice for periodic crystals. The expression "satellites" (usually with weaker intensity with respect to the main reflections) was used to describe those reflections for which the fourth index is non-zero. The key to interpreting the structure was the concept of *superspace*, a higher dimensional (in this case, four-dimensional) space [6]. The incommensurate or modulated structure in three-dimensions can be viewed as the projection of a periodic four-dimensional superspace structure into three dimensions (e.g., [7]).

Mineralogists discovered other complex materials that challenged the hegemony of periodicity. For example, Makovicky and Hyde in their landmark paper of 1981 [8] reported several mineral "layered" structures that consist of two periodic interpenetrating components that are incommensurate in the direction parallel to the layers and commensurate normal to the layers, thus introducing the concept of incommensurate (misfit) layer structures into condensed matter physics. Structures with similar features are currently known as *composite* structures. It was later realized that composite structures could also be described in using a higher-dimensional superspace.

The discovery of incommensurately modulated and composite structures shook the foundations of crystallography but did not break them. All the examples obeyed the same rotational symmetry restrictions as conventional crystals. Also, the incommensurability was not absolutely fixed. Even small variations in temperature and pressure can alter the modulation, transforming incommensurate irrational ratios of periods into rational ones in some cases.

4.2 Revolution in Crystallography: Quasicrystals (1984)

The great departure from periodicity and the structures of conventional crystallography was the introduction of the concept of quasicrystals in 1984 by Dov Levine and one of us (PJS) [9]. Quasicrystals, by definition, have an atomic arrangement that is quasiperiodic – two or more types of atomic clusters each repeat at regular intervals whose ratio is irrational (not expressible as a fraction). ("Quasicrystal" is short for "quasiperiodic crystal"). Unlike a periodic crystal, the sequence of atoms never exactly repeats in the sense that no two atoms have exactly same arrangements of atoms surrounding them (if the atomic arrangement is followed sufficiently far out). Levine and Steinhardt showed that quasiperiodic structures can violate the mathematical constraints of conventional crystallography and exhibit any rotational symmetry forbidden to crystals, including five-fold symmetry in the plane and icosahedral symmetry

Fig. 4.1: (a) Fragment of a two-dimensional Penrose tiling composed of two types of tiles arranged quasiperiodically with crystallographically-forbidden five-fold symmetry; (b) Fragment of a three-dimensional icosahedral quasicrystal composed of four types of polyhedral units with holes and protrusions that constrain the way the units match face-to-face in such a way as to guarantee that all space-filling arrangements are quasicrystalline. For details of the construction, see Socolar and Steinhardt [33]. Figure after Bindi and Steinhardt [27].

(the symmetry of a soccer ball) in three dimensions. An inspiration for the idea was the five-fold symmetric Penrose tiling composed of two tile shapes that repeat with frequencies whose ratio is the famous irrational number known as the "golden ratio" (Fig. 4.1a). Unlike the case of incommensurate crystals, the irrational ratios of periods are fixed by the non-crystallographic symmetry and do not shift continuously with changes in pressure and temperature.

In 1984, Dan Shechtman, Ilan Blech, Denis Gratias and John Cahn [10], reported the discovery of an aluminum-manganese alloy that has point-like electron diffraction pattern like a crystal but with forbidden icosahedral symmetry. Levine and Steinhardt had not yet publicized their idea, so the Shechtman et al. paper did not consider it or propose any theory to explain their profound rule-violating discovery. Levine and Steinhardt showed that the diffraction pattern they observed closely matched the one computed for an icosahedral quasicrystal and proposed that Shechtman's discovery may be an example of the quasicrystal concept.

Notably, ideas originally developed for the earlier incommensurately modulated crystals can be generalized for quasicrystals. For example, the diffraction pattern for an icosahedral quasicrystal can be indexed using six rational indices, and the real space structure can be viewed as a three-dimensional slice through a six-dimensional periodic superspace structure. The superspace picture is now a common approach for describing all known *aperiodic* structures, including incommensurately modulated, composite and quasicrystalline structures, although other types of ordered structures may be possible that cannot submit to a superspace description.

The original announcements by Shechtman et al. and Levine and Steinhardt were greeted with skepticism because the experimental diffraction pattern was not quite point-like and the reflections were not quite ideally located compared to the ideal quasicrystal picture. Also, it was conjectured that atoms could not self-organize into a non-repeating pattern, like a quasicrystal, without introducing a high density of defects. A turning point was the successful synthesis of the $Al_{63}Cu_{24}Fe_{13}$ quasicrystal by An-Pang Tsai and collaborators in 1987 [11]. This alloy exhibited high resolution-limited Bragg peaks and an extremely high degree of structural perfection clearly fitting the quasicrystal hypothesis. Competing models, such as the Shechtman-Blech random cluster picture, the Stephens-Goldman icosahedral glass picture, and the Pauling multiply-twinned picture, could not explain the new diffraction patterns [12]. In this sense, $Al_{63}Cu_{24}Fe_{13}$ might be considered the first *bona fide* quasicrystal.

Today, despite the increasing number of different quasicrystal alloys synthesized in the laboratory and careful analysis (e.g., [13, 14]), there is still not a general consensus about the status of quasicrystals as a fundamental state of matter. Some believe that quasicrystals can be energetically stable states of matter in which atomic clusters join together in energetically favorable configurations analogous to the way Penrose tiles fit together according to matching rules into a perfect quasiperiodic pattern, as conjectured by Levine and Steinhardt [9]. Others argue that quasicrystals are, at best, entropically stable analogous to random tilings (Penrose-like tiles without matching rules randomly arranged to fill a space) with icosahedral symmetry and quasiperiodicity appearing for entropic reasons when averaging over large configurations [15]. According to this picture, quasicrystals are inherently delicate, metastable oddities that may only be synthesized for ideal compositions and under highly controlled laboratory conditions. On the contrary, the original concept [construction using polyhedral units and certain "face-to-face matching" rules that force a quasiperiodic arrangement with perfect long-range icosahedral order (Fig. 4.1b)] considered quasicrystals to be energetically stable and robust as crystals and to form even in non-ideal conditions.

4.3 Revolution in Mineralogy: Natural Quasicrystals (2009)

The search for quasicrystals in nature was motivated in large part as an attempt to shed light on the quasicrystal debate. If quasicrystals are energetically stable and robust as crystals, then it is conceivable that they formed under natural conditions, just like crystalline minerals. Although recent extensive studies in the laboratory by Tsai and others (for a comprehensive review see [16]) have shown that quasicrystals are almost certainly stable states of matter in some cases, as Levine and Steinhardt [9] long argued, rather than a random, decomposing aggregate, the discovery of an ancient (million years old) natural example, would make the case even more compelling. One of us (PJS) began to consider this possibility soon after the first papers on quasicrystals appeared.

The first attempts to find natural quasicrystals involved casually looking for candidates whenever visiting extant museum mineral displays, but nothing came of this approach. Then, in 1998, an idea for searching for natural quasicrystals systematically was conceived that took advantage of the existence of computer databases of X-ray powder diffraction patterns that could be accessed online. X-ray powder diffraction patterns are obtained by scattering X-rays from a mixture of many grains oriented at random angles such that the diffraction pattern is circularly symmetric. Powder patterns have the disadvantage that they obscure the rotational symmetry. On the other hand, a computer database containing the powder diffraction patterns of over eighty thousand samples including nine thousand mineral samples maintained by the International Center for Diffraction Data represents a rich source of information to tap if one knows what to do with it. Working with Peter Lu, Ken Deffeyes and Nan Yao, a set of quantitative measures was devised that could identify the most promising quasicrystal candidates in the database, including a handful of minerals. Each of the minerals was acquired and studied by transmission electron microscopy (TEM) and single-crystal X-ray diffraction to check for forbidden symmetry, but, in the end, no new quasicrystals, synthetic or natural, were discovered in the original study. The results were published in a paper [17], which included at the end an offer to share the names of additional candidates on the list with any collaborators willing to test minerals from their collection.

Six years later, the call was answered by one of us (LB) who began by testing samples on the list that were also in the collections of the Museo di Storia Naturale of the Università degli Studi di Firenze (Italy). When that effort failed, Bindi suggested examining a sample labelled “khatyrkite” (catalog number 46407/G; Fig. 4.2), a metallic crystal phase of copper and aluminum, $(Cu,Zn)Al_2$, based on the fact that many known synthetic quasicrystals include metallic aluminum. According to the museum catalog, the sample originated from the Khatyrka region of the Koryak mountains in the Chukotka autonomous okrug on the north-eastern part of the Kamchatka peninsula [18–20]. The first report of khatyrkite appeared in a paper by Razin et al. [21] in 1985. According to that report, khatyrkite is a tetragonal mineral found in association with cupalite, nominally (Cu,Zn)Al, an orthorhombic phase. The Florence sample did not consist purely of metal, though. Other typical rock-forming minerals [e.g., forsterite, $(Mg,Fe)_2SiO_4$, diopside, $CaMgSi_2O_6$, spinel $MgAl_2O_4$] were intergrown with the metals, which included cupalite, β-AlCuFe and, most significantly, a few grains of a new species, whose composition $Al_{63}Cu_{24}Fe_{13}$ had not been found previously in any mineral [18, 19]. Intriguingly, the composition was close to that found by Tsai et al. [11] for a synthetic quasicrystal.

Initial tests of X-ray powder pattern produced promising results, but other promising powder patterns had found early in the search for candidates that later proved to be duds. The key test was obtained with a transmission electron microscope from iso-

Fig. 4.2: Different views of the original khatyrkite-bearing sample belonging to the collections of the Museo di Storia Naturale of the Università degli Studi di Firenze (catalogue number 46407/G). The lighter-coloured material on the exterior contains a mixture of spinel, clinopyroxene and olivine. The dark material consists predominantly of khatyrkite ($CuAl_2$) and cupalite (CuAl), but also includes granules of icosahedrite with composition $Al_{63}Cu_{24}Fe_{13}$. Figure after Bindi and Steinhardt [27].

lated tiny grains extracted from the sample. The sample was sent to Princeton University where the test was performed at 7:00 am on the very first day of 2009. The results were dramatic and clear: the diffraction patterns of the natural $Al_{63}Cu_{24}Fe_{13}$ grains had the precise signature of an icosahedral quasicrystal (Fig. 4.3)! Self-similar patterns of sharp peaks arranged in straight lines with five-, three- and two-fold symmetry were observed along different axes separated by the characteristic angles of an icosahedron [18].

The degree of perfection was especially impressive. They exceeded the results obtained in the original aluminum-manganese phase found by Shechtman et al. [10] or in typical synthetic quasicrystals. Quasicrystals made by rapid quenching and/or embedded in a matrix of another phase generically exhibit easily detectable phason strains [22, 23]. Phasons are hydrodynamic (gapless) modes that occur in incommensurate crystals and quasicrystals in addition to the usual phonon modes. If the atomic density $\rho(x)$ is decomposed into a sum of incommensurate periodic density waves, a phason *shift* corresponds to a uniform phase shift between density waves whose periods have an irrational ratio. A phason strain corresponds to a spatial gradient in the

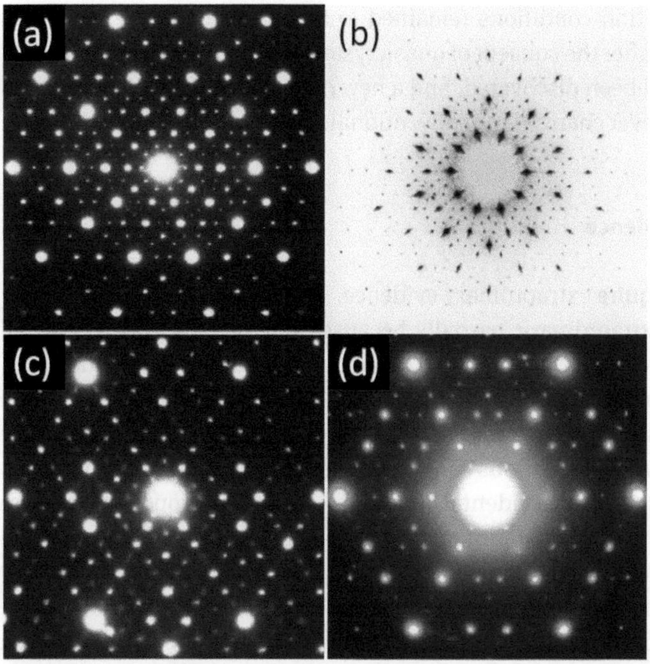

Fig. 4.3: The unmistakable signature of an icosahedral quasicrystal consists of planar patterns of sharp peaks arranged along straight lines in an incommensurate lattice with five-fold (a), two-fold (c) and three-fold (d) symmetry depending on the beam direction. The electron diffraction patterns shown here, taken from a grain of icosahedrite, match those predicted for a face centered icosahedral quasicrystal, as do the angles that separate the symmetry axes. In panel (b) is reported a single-crystal X-ray diffraction precession photograph down one of the five-fold symmetry axes of the icosahedron collected on a small icosahedrite fragment removed from a grain recovered in the Koryak expedition. Figure after Bindi and Steinhardt [27].

phase shift. At the atomic level, phason strains produce rearrangements of atoms relative to the ideal ground-state configuration that require diffusive motion to relax away, so the phason strains remain after solidification unless the sample is grown using a very carefully-controlled protocol. In reciprocal space, the signature of phason strain is a systematic shift in the Bragg peak positions from the ideal by an amount that increases as the peak intensity decreases. The effect is easily observed by holding the diffraction pattern at a grazing angle and viewing along rows of peaks. The phason strain can be observed as deviations of the dimmer peaks from straight lines [23].

Yet, no detectable phason strain can be found in the electron diffraction patterns in Fig. 4.3 despite the fact that icosahedrite was not grown under controlled laboratory conditions. Instead, this degree of perfection was obtained in a quasicrystal intergrown with other phases under conditions that are clearly far from equilibrium (Fig. 4.2). Somehow Nature had managed to accomplish a feat that humans had to work very hard to achieve.

Although the formation conditions remained to be understood, one point was clear: Twenty-five years after the concept of quasicrystals was first introduced, the first natural quasicrystal had been discovered, and a new chapter in the study of mineralogy had opened that forever changed the conventional classification of mineral forms.

4.3.1 Extraordinary evidence

Extraordinary claims require extraordinary evidence. The claimed discovery of a natural quasicrystal was extraordinary, partially because none had ever been detected before but especially because of the presence of metallic aluminum in the quasicrystal grains as well as some of the other mineral phases found in the rock. Aluminum normally oxidizes unless placed in a highly reducing environment, as is done in the laboratory or in industry. Hence, serious consideration had to be given to the possibility that the sample was slag or an accidental byproduct of some anthropogenic process (including rocket or plane exhaust) or intentional fakery. Extraordinary evidence had to be assembled to address these concerns.

At first, there was no direct evidence where the Florence sample actually came from other than the marking on the sample box indicating the source as the Koryak Mountains of the Kamchatka Peninsula in far eastern Russia. Only through a remarkable investigation stretching over a year and worthy of a detective novel was the claim ultimately confirmed. Ultimately, the history of the sample was traced back to the person, Valery V. Kryachko, who originally unearthed it from a blue-green clay bed along the Listventovyi stream off the Iomrautvaam tributary of the Khatyrka River in the Koryak Mountains in 1979 [24].

As this detective story was being pursued, more laboratory data was collected from the Florence sample that provided overwhelming evidence that the rock was formed by a natural process. Particularly telling was the discovery of a grain of quasicrystal trapped within stishovite (Figs. 4.4 and 4.5), a polymorph of SiO_2 that only forms at ultrahigh pressures (>100,000 times atmospheric pressure) [20]. Any concerns that the sample was slag or a human-made hoax were now eliminated. Of all the explanations that had been considered, the only two remaining possibilities were that the quasicrystal formed near the core-mantle boundary and was projected to the surface in a superplume, or that it formed in a violent collision between meteorites in space.

The evidence for naturalness was now sufficient for the International Mineralogical Association to accept the new mineral as the first natural quasicrystal to be entered into its official catalog. The mineral was assigned the name *icosahedrite,* in recognition of the icosahedral symmetry of its atomic structure [19]. The sample is identified as having a face-centred icosahedral symmetry (abbreviated as $Fm\overline{35}$) and its diffraction pattern is characterized by six Miller-like indices corresponding to the six basis vectors that define an icosahedral reciprocal lattice.

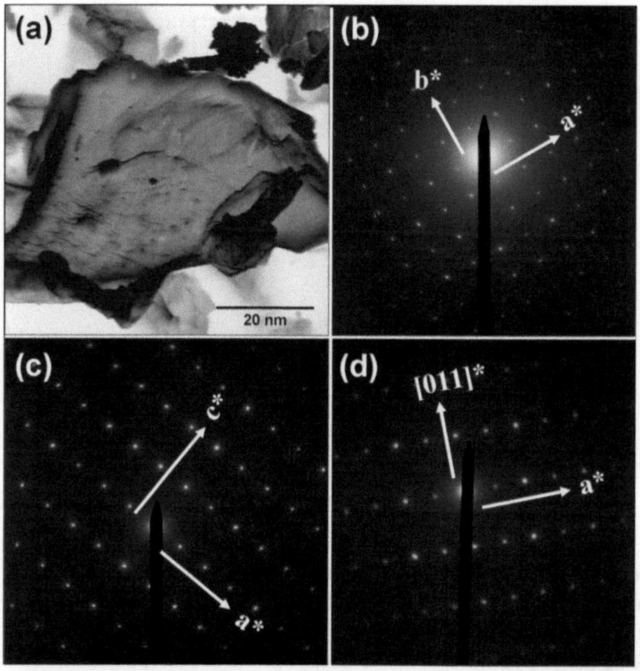

Fig. 4.4: (a) TEM image of a 50 nm grain of stishovite, a tetragonal polymorph of SiO_2; (b) selected area diffraction pattern taken down the [001] axis; (c) selected area diffraction pattern taken down the [010] axis; (d) selected area diffraction pattern taken down the [0–11] axis. All the collected diffraction patterns establish the tetragonal symmetry and lattice parameters $a = 4.2$ Å and $c = 2.7$ Å, typical for the mineral stishovite. Figure after Bindi et al. [20].

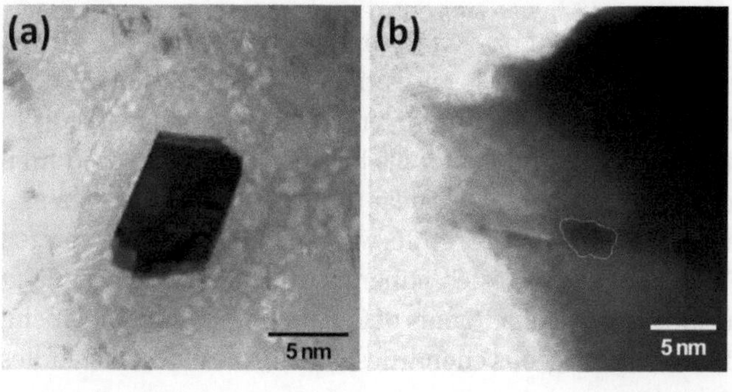

Fig. 4.5: Panel (a) shows an inclusion of icosahedrite in the stishovite grain shown in Fig. 4a; (b) Rotated TEM image showing enclosed icosahedrite which, when combined with (a), establishes that icosahedrite is totally encased in stishovite (the boundary of inclusion has been depicted with a dashed line for clarity). Figure after Steinhardt and Bindi [24].

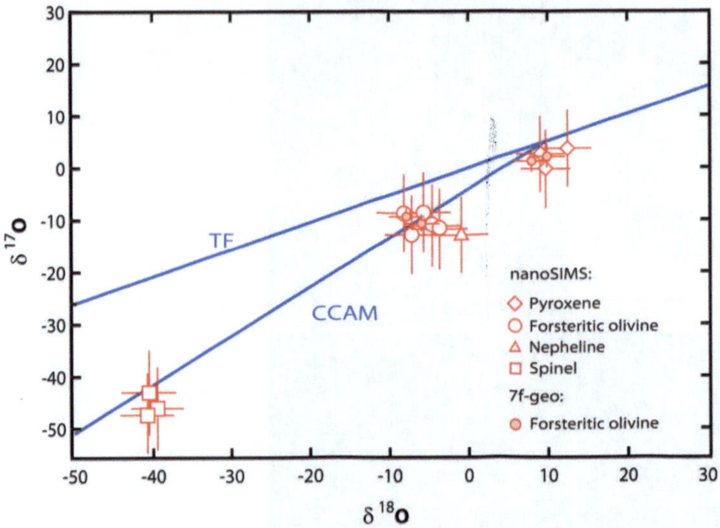

Fig. 4.6: A plot of oxygen three-isotope composition as suggested by Clayton et al. [25] to discriminate terrestrial and extra-terrestrial minerals, containing data for four minerals (pyroxene, nepheline, olivine, and spinel) in the sample studied. All data are ion probe measurements made with either the Cameca nanoSIMS (large error bars) or the Cameca 7f geo (small error bars). The error bars are 2σ. The plotted compositional indices, $\delta^{18}O_{VSMOW}$ and $\delta^{17}O_{VSMOW}$, are respectively the difference between the ratios $^{18}O/^{16}O$ and $^{17}O/^{16}O$ and the same ratios in Vienna Standard Mean Ocean Water (VSMOW), expressed in parts per mil. Terrestrial minerals fall along the upper gently sloping line (TF); the oxygen isotope compositions measured for our sample lie along the line corresponding to anhydrous minerals in the CO or CV carbonaceous chondrites (the 'CCAM'). Figure after Bindi et al. [20].

The next step was to measure the ratio of oxygen isotopes ($^{18}O/^{16}O$ and $^{17}O/^{16}O$) to determine if origin of the Florence sample was terrestrial or extraterrestrial [25]. Anhydrous carbonaceous chondrites commonly contain minerals that differ markedly in oxygen isotope composition from terrestrial materials because they condensed from or underwent isotopic exchange with nebular gas near the formation of the solar system that had a different radiogenic history. The results were unambiguous: the oxygen isotope measurements (Fig. 4.6) of all the tested minerals in the Florence sample fit tightly the characteristic results for the constituents of Calcium Aluminum rich Inclusions (CAIs) from CV3 and CO3 carbonaceous chondrites, among the oldest meteorites to have formed in our solar system.

To go the next step and understand how the copper-aluminum alloys formed in the early solar system required more than the tiny specks of powder that remained from the Florence sample after all the studies. New extraordinary evidence was needed, and the only way to get that was to return the place where the Florence sample was first found.

Despite the difficulties and the unlikely possibility of finding more samples, a geological expedition to Chukotka in far eastern Russia was organized and undertaken in 2011 [24, 26, 27]. The journey to one of the most remote places on the planet (Figs. 4.7 and 4.8) resulted in the unexpected discovery of new nine samples [27, 28] whose properties have firmly established that the quasicrystal and the rock containing it are part of a carbonaceous chondritic meteorite (now officially named *Khatyrka*, after the river in Chukotka; see [28]) with calcium aluminum inclusions that date back 4.5 Gya to the formation of the solar system.

Fig. 4.7: Members of the Koryak expedition team (left to right): Will Steinhardt (Harvard, USA), Vadim Distler (IGEM, Russia), Luca Bindi (University of Firenze, Italy), Alexander Kostin (BHP Billiton, USA), Michael Eddy (MIT, USA), Christopher Andronicos (Cornell, USA), Glenn MacPherson (Smithsonian Institution, USA), Valery Kryachko (Voronezh, IGEM), Paul Steinhardt (Princeton, USA) and Marina Yudovskaya (IGEM, Russia).

Some of the recovered grains show petrographic features clearly indicating that the meteorite underwent an impact shock that produced a heterogeneous distribution of pressures and temperatures in which some portions reached at least 5 GPa and 1200 °C [29]. The conditions would have been sufficient to melt Al-Cu-bearing minerals, which then rapidly solidified into icosahedrite and other Al-Cu metal phases. The meteorite also contains heretofore unobserved phases of iron-nickel and iron sulfide with substantial amounts of Al and Cu, including the recently described new mineral steinhardtite [30].

The abundant and consistent information gained from tracking down the source of the Florence sample, conducting the expedition to the Koryaks, and finding grains

Fig. 4.8: Field operations at the Listventovyi Stream (clockwise): extracting clay from areas along the stream, panning the clay down to mineral separates, mapping the structural geology and examining mineral samples.

containing new mineral phases and evidence of high-velocity impact shock unobserved previously in other CV3 carbonaceous chondrites constitutes the "extraordinary evidence" needed to show that icosahedrite is the result of natural processes that occurred in the early solar system.

4.3.2 Impact

The discovery of the first natural quasicrystal is a breakthrough in mineralogy that may be just the beginning of a new era. Could there possibly be other natural quasicrystals to be discovered? As it turns out, the answer to that question is already known. Recently, Bindi et al. [31] reported a phase of iron-nickel with a much higher fraction of aluminum than crystalline steinhardtite. The new phase, with composition $Al_{71}Ni_{24}Fe_5$, is the second natural quasicrystal to be found (Fig. 4.9). The find is historic because it is the first example of a natural quasicrystal with decagonal symmetry (ten-fold symmetry in the plane and periodic along the axis perpendicular to the plane). The new quasicrystal was found associated with steinhardtite ($Al_{38}Ni_{32}Fe_{30}$), Fe-poor steinhardtite ($Al_{50}Ni_{40}Fe_{10}$), Al-bearing trevorite ($NiFe_2O_4$) and Al-bearing taenite (FeNi). As it was the case for icosahedrite, the new decagonal phase had been synthesized in the laboratory previously. The laboratory studies showed that it is stable over a narrow range of temperatures, 1120K to 1200K at standard pressure, adding further support for the conclusion that the Khatyrka meteorite reached heterogeneous high temperatures [$1100 < T(\mathrm{K}) \leq 1500$] and then rapidly cooled after being heated during an impact-induced shock.

These mineralogical discoveries of new quasicrystals have an impact on other scientific disciplines. Icosahedrite, the new decagonal quasicrystal phase, and all the aluminum-bearing crystalline alloys present a challenge for meteorite science and our understanding of novel processes in the early solar system. We still are not certain of how the natural quasicrystals in the Khatyrka meteorite formed. At present, two possible scenarios have been described in the literature: (*i*) the Al-bearing FeNi phases might have been the initial source of aluminum in the Al-bearing alloys khatyrkite, cupalite and icosahedrite; or (*ii*) the Al-metals may have had a pre-accretion nebular origin and steinhardtite and Al-bearing taenite observed in the sample formed by reaction of shock-produced Al-melt and pre-existing taenite. In the first hypothesis, the shock features observed in parts of some grains of the Khatyrka meteorite [29] would be generated by a strong increase of heat and pressure sufficient to extract Al from the FeNi metals and to initiate the local melting of metals and silicates. In the second hypothesis, the Al metals would form in some nebular process before the impact, with the impact resulting in the remelting, rapid cooling (about 10^2–10^3 °C s^{-1}) and solidification of the Al metals. (Other hypotheses for naturally producing metallic Al are also being considered and tested presently.) In any scenario, the sequence of events leading to the exchange of metallic Al that formed steinhardtite, the new decagonal quasicrystal and Al-taenite can only be plausibly imagined to occur in space under low fO_2 solar nebular conditions.

Icosahedrite and the crystalline alloys that contain both Al and Cu present a challenge to geochemistry. As is well known, in the early stages of the solar system, aluminum formed solids when copper was still a gas. Also, aluminum has an affinity for oxygen and copper an affinity for sulfur. Understanding the formation of the Al-Cu alloys in Khatyrka could provide insights about a spectrum of geochemical processes that were unknown before.

The search for natural quasicrystals has raised numerous interesting issues for those interested in the evolution of the solar system. It seems clear now that, even if we do not understand how, quasicrystals formed in the early stages of the solar system that ultimately led to the formation of planets. This places quasicrystals among the first 100 minerals known to have formed in our solar system. It is too early to quantify the relative abundance, but the fact that examples have been seen at all suggests

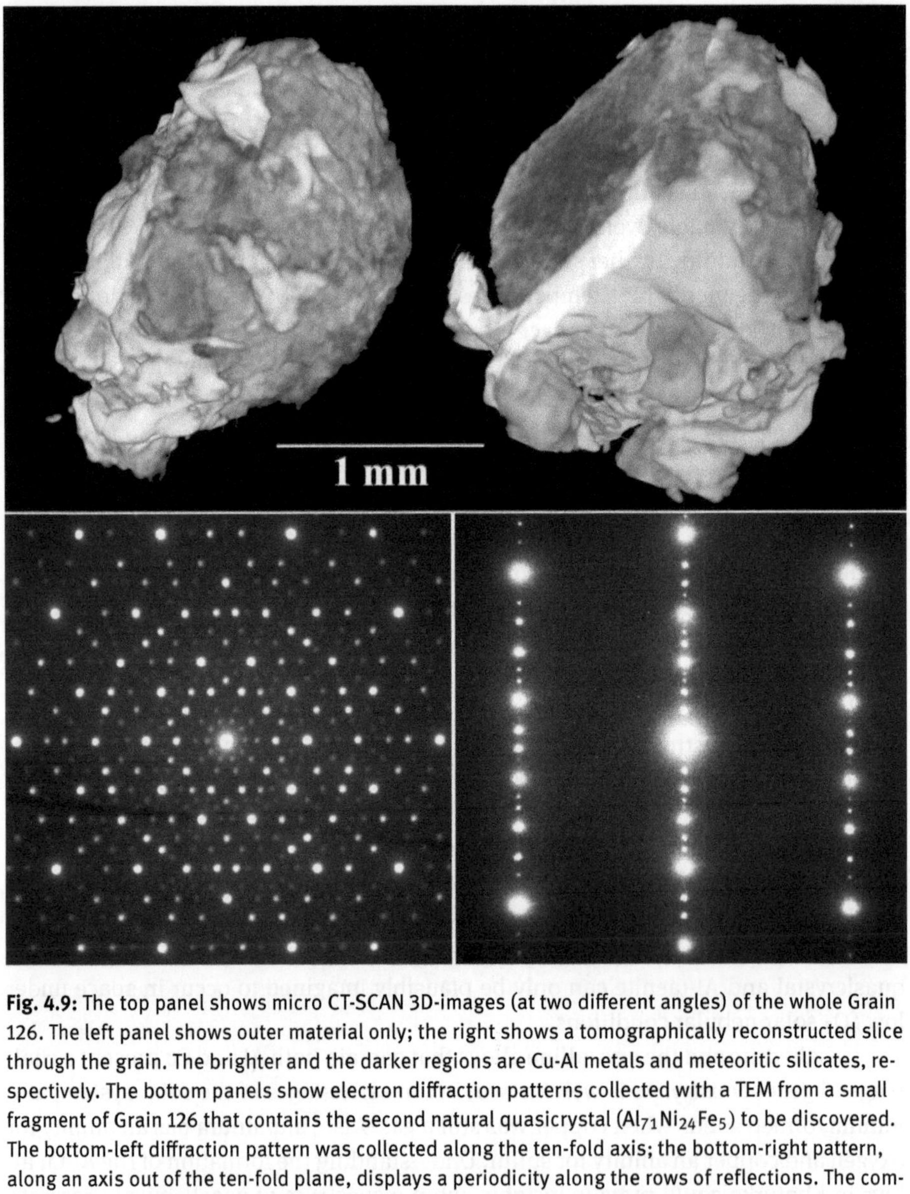

Fig. 4.9: The top panel shows micro CT-SCAN 3D-images (at two different angles) of the whole Grain 126. The left panel shows outer material only; the right shows a tomographically reconstructed slice through the grain. The brighter and the darker regions are Cu-Al metals and meteoritic silicates, respectively. The bottom panels show electron diffraction patterns collected with a TEM from a small fragment of Grain 126 that contains the second natural quasicrystal ($Al_{71}Ni_{24}Fe_5$) to be discovered. The bottom-left diffraction pattern was collected along the ten-fold axis; the bottom-right pattern, along an axis out of the ten-fold plane, displays a periodicity along the rows of reflections. The combination of quasiperiodicity and ten-fold symmetry in one plane and periodicity along the third dimension is characteristic of decagonal symmetry. (By contrast, icosahedrite is quasiperiodic along all symmetry axes.)

that it cannot be negligible. Further, if quasicrystals formed in our solar system, it is reasonable to suppose that it formed in other solar systems in the Milky Way and in galaxies throughout the cosmos. Conceivably, quasicrystals may have also formed in the highly reducing conditions near the core-mantle boundary, as speculated during the early stages of the investigation when we did not yet know that the Florence sample was extraterrestrial. This possibility seems worthy of exploring since it may give us new insights on core composition and properties. More generally, the discovery of the first quasicrystals should trigger the re-examination of other terrestrial and extraterrestrial minerals in search of different quasicrystals.

From the perspective of condensed matter physics, the observations that the samples formed under astrophysical conditions constitute significant new support for the original proposal that quasicrystals can be energetically stable states of matter, on the same footing as crystals [9]. The alternative entropic picture, which assumes that the atoms have enough time to explore many different random configurations, is hard to reconcile with the extreme conditions under which icosahedrite and the new decagonal quasicrystal formed and annealed with a high degree of structural perfection. The finding of natural quasicrystal enclosed within stishovite and other evidence which indicates that formation likely occurred under shock conditions and rapid cooling where entropic effects were unimportant compared to energetics and kinetics in forming the phase [29]. Furthermore, all the icosahedrite grains that have been studied thus far have the same composition, to within better than one per cent, even though the phases surrounding them are different. This observation is expected if the composition corresponds to an energetically preferred structure, whereas the entropic model generally allows a continuous range of composition (corresponding to crystal approximant structures). Further laboratory studies could shed light on this issue. For example, our recent investigations of synthetic $Al_{63}Cu_{24}Fe_{13}$ by means of *in situ* synchrotron X-ray diffraction experiments using multianvil device clearly show that it is stable up to ~5 GPa and 1673 K with a negligible effect of pressure on the volumetric thermal expansion properties [32]. In addition, the structural analysis of the recovered sample has excluded the transformation of the AlCuFe quasicrystalline phase to possible approximant phases, which is in contrast with previous predictions at ambient pressure. Results from our preliminary study [32] extended the knowledge on the stability of icosahedral AlCuFe at higher temperature and pressure than previously examined and have provided a new constraint on the stability of icosahedrite. Moreover, we are currently working on the phase diagram of AlCuFe-quasicrystals up to very high pressures to better understand the stability of icosahedrite in nature.

4.4 Outlook

The birth of mineralogy as a quantitative science, as pioneered by Haüy, led to profound new insights across the physical sciences. The exploration of natural quasicrystals described here demonstrates that mineralogy can continue to surprise us and can have an impact on other disciplines, including geoscience, solar system evolution, planet formation, condensed matter physics, and materials engineering.

From the physics point of view, perhaps the greatest puzzle is to understand why quasicrystals are more stable than crystals for some compositions. Further explorations in mineralogy may also lead us to discover quasicrystal compositions and symmetries never observed previously, leading to new materials with novel physical properties of some utility.

Acknowledgment: This paper reviews work that has benefitted greatly from our valued collaborators K. Deffeyes, J.M. Eiler, Y. Guan, L.S. Hollister, C. Lin, P.J. Lu, J. Poirier, N. Yao, and the expeditionary team composed by C. Andronicos, V.V. Distler, M.P. Eddy, V. Komelkov, O. Komelkova, A. Kostin, V. Kryachko, G.J. Macpherson, B. Makovskii, W. Steinhardt, and M. Yudovskaya. The expedition was supported through a contribution to Princeton University by an anonymous donor, to whom we are most grateful. Many who have given valuable advice and support including C. Ballhaus, J. Beckett, P. Bonazzi, A. El Goresy, A. Ishiwatari, H.-C. Jeong, J. Jones, S. Menchetti, M. Morozov, G. Poirier, P. Robinson, V. Rudashevkij, P. Spry, and E. Stolper. We give special thanks to the IGEM (Moscow) for their support in organizing the expedition to Chukotka and collaboration. This work is supported in part by the National Science Foundation Materials Research Science and Engineering Center program through New York University Grant DMR-0820341.

Bibliography

[1] Penfield SL, Ford WE. Ueber den Calaverit. Zeit Kristallogr 1902, 35, 430–451.

[2] Smith H. Ueber das bemerkenswerthe Problem der Entwicklung der Krystallformen des Calaverit. Zeit Kristallogr 1903, 37, 209–234.

[3] Goldschmidt V, Palache C, Peacock M. Ueber Calaverit. N Jahr Mineral Monat 1931, 63, 1–58.

[4] Dam B, Janner A, Donnay JDH. Incommensurate morphology of calaverite ($AuTe_2$) crystals. Phys Rev Lett 1985, 55, 2301–2304.

[5] Janner A, Dam B. The morphology of calaverite ($AuTe_2$) from data of 1931 – solution of an old problem of rational indexes. Acta Crystallogr 1989, A45, 115–123.

[6] Janssen T, Janner A, Looijenga-Vos A, Wolff de PM. Incommensurate and commensurate modulated structures. International Tables for X-ray Crystallography, Vol. C, 899–937, 1992, Kluwer Academic Publishers.

[7] Chapuis G. Crystallographic excursion in superspace. Crystal Engin 2003, 6, 187–195.

[8] Makovicky E, Hyde BG. Non-commensurate (misfit) layer structures. Inorg Chem 1981, 46, 101–170.

[9] Levine D, Steinhardt PJ. Quasicrystals: A new class of ordered structures. Phys Rev Lett 1984, 53, 2477–2480.

[10] Shechtman D, Blech I, Gratias D, Cahn JW. Metallic Phase with Long-Range orientational order and no translational symmetry. Phys Rev Lett 1984, 53, 1951–1953.

[11] Tsai A-P, Inoue A, Masumoto T. A Stable Quasicrystal in Al-Cu-Fe System. Jap J Appl Phys 1987, 26, L1505.

[12] Bancel PA, Heiney PA, Horn PM, Steinhardt PJ. Comment on a Paper by Linus Pauling. Proc Natl Acad Sci USA 1989, 86, 8600–8601.

[13] Janot C. Quasicrystals: A Primer. DiVincenzo D and Steinhardt PJ Eds, Oxford University Press, 1997, Oxford.
[14] Steurer W, Deloudi S. Crystallography of Quasicrystals. Concepts, Methods and Structures. Springer, 2009, Berlin.
[15] Henley C. In Quasicrystals: The State of the Art, eds DiVincenzo D & Steinhardt PJ (World Scientific, Singapore), 1991, 429–524.
[16] Tsai A-P. Discovery of stable icosahedral quasicrystals: progress in understanding structure and properties. Chem Soc Rev 2013, 42, 5352–5365.
[17] Lu PJ, Deffeyes K, Steinhardt PJ, Yao N. Identifying and indexing icosahedral quasicrystals from powder diffraction patterns. Phys Rev Lett 2001, 87, 275507.
[18] Bindi L, Steinhardt PJ, Yao N, Lu PJ. Natural quasicrystals. Science 2009, 324, 1306–1309.
[19] Bindi L, Steinhardt PJ, Yao N, Lu PJ. Icosahedrite, $Al_{63}Cu_{24}Fe_{13}$, the first natural quasicrystal. Am Mineral 2011, 96, 928–931.
[20] Bindi L, Eiler JM, Guan Y et al. Evidence for the extra-terrestrial origin of a natural quasicrystal. Proc Natl Acad Sci USA 2012, 109, 1396–1401.
[21] Razin LV, Rudashevskij NS, Vyalsov LN. New natural intermetallic compounds of aluminum, copper and zinc – khatyrkite $CuAl_2$, cupalite CuAl and zinc aluminides from hypui erbasites of dunite-harzburgite formation. Zap Vses Mineralog Obshch 1985, 114, 90–100 (in Russian).
[22] Levine D, Lubensky TC, Ostlund S, Ramaswamy A, Steinhardt PJ. Elasticity and defects in pentagonal and icosahedral quasicrystals. Phys Rev Lett 1985, 54, 1520–1523.
[23] Lubensky TC, Socolar JES, Steinhardt PJ, Bancel PA, Heiney PA. Distortion and peak broadening in quasicrystal diffraction patterns. Phys Rev Lett 1986, 57, 1440–1443.
[24] Steinhardt PJ, Bindi L. In search of natural quasicrystals. Rep Progr Phys 2012, 75, 092601–092611.
[25] Clayton RN, Onuma N., Mayeda TK. A classification of meteorites based on oxygen isotopes. Earth Plan Science Lett 1976, 30, 10–18.
[26] Bindi L, Steinhardt PJ. The discovery of the first natural quasicrystal. A new era for mineralogy? Elements 2012, 8, 13–14.
[27] Bindi L, Steinhardt PJ. The quest for forbidden crystals. Mineral Mag 2014, 78, 467–482.
[28] MacPherson GJ, Andronicos C, Bindi L et al. Khatyrka, a new CV3 find from the Koryak Mountains, Eastern Russia. Met Plan Sci 2013, 8, 1499–1514.
[29] Hollister LS, Bindi L, Yao N. et al. Impact-induced shock and the formation of natural quasicrystals in the early solar system. Nat Comms 2014, 5, 3040.
[30] Bindi L, Yao N, Lin C, et al. Steinhardtite, a new body-centered-cubic allotropic form of aluminum from the Khatyrka CV3 carbonaceous chondrite. Am Mineral 2014, 99, 2433–2436.
[31] Bindi L, Yao N, Lin C, et al. Natural quasicrystal with decagonal symmetry. Sci Rep 2015, 5, 9111.
[32] Stagno V, Bindi L, Shibazaki Y et al. Icosahedral AlCuFe quasicrystal at high pressure and temperature and its implications for the stability of icosahedrite. Sci Rep 2014, 4, 5869.
[33] Socolar J, Steinhardt PJ. Quasicrystals II: Unit cell configurations. Phys Rev B 1986, 34, 617–647.

Fabrizio Nestola

5 Ringwoodite: its importance in Earth Sciences

5.1 History of ringwoodite

The history of ringwoodite started in 1869 in a remote locality in the south-west of Queensland in Australia. Mr. Michael Hammond witnessed a meteorite shower close to the junction between Cooper and Kyabra Creeks (Lat. 25° 30′ S., Long. 142° 40′ E.), not far from Windorah (Queensland, Australia) and about 1000 km west of Brisbane. The meteorite fall was very impressive and in due course 102 stones were recovered. Mr. Hammond was the owner of the Tenham Station and from this the meteorite collection was named as "Tenham meteorites". This collection was then offered in 1935 to the British Museum by Mr. Benjamin Dunstan, formerly Government Geologist of Queensland [1].

But why does this nice story match with ringwoodite? In 1969, exactly 100 years after Mr. Hammond observed the Tenham meteorite fall, R. A. Binns, R. J. Davies and S. J. B. Reed published in Nature [2] the first natural evidence of ringwoodite after studying a fragment of the Tenham meteorite. Thirty years later Chen et al. [3] reported clear images of some lamellae of about 1–2 μ in thickness showing a higher density than olivine but with identical composition (Fig. 5.1, modified from Chen et al. [3]). The

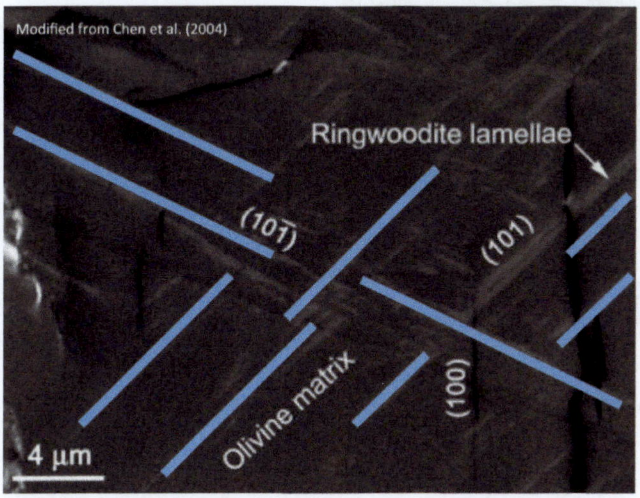

Fig. 5.1: Back-scattered image of lamellae of ringwoodite in olivine (modified from [3]). The lamellae are evident being marked by a brighter grey. The darker grey corresponds to olivine. The blue solid lines are reported to indicate the directions along which ringwoodite grew. In white parentheses the lattice planes are indicated. The thickness of the lamellae could be on the sub-μm scale.

X-ray powder diffraction data of the same lamellae inside olivine [2] indicated cubic symmetry with the cell edge $a = 8.113(3)$ Å. This discovery showed that this new cubic polymorph of olivine was identical to the phase synthesized three years earlier [4] at about 17 GPa and a temperature estimated to be around 900 °C. This cubic polymorph was obtained at experimental conditions typical of deep mantle, demonstrating that this phase could be a major component of this layer of our planet. Thus, a proposal for a new mineral was submitted [2] to the International Mineralogical Association (IMA) leading to approval of this new cubic polymorph of olivine as RINGWOODITE (IMA 1969-038). The name was given in honor of Prof. Alfred Edward Ringwood of the Australian National University for his experimental studies on petrology, phase transformations, constitution, and dynamics of the mantle.

The stability field of ringwoodite falls within the *P-T* conditions of the so-called transition zone. The transition zone extends between 410 and 660 km depth. It does not show any complex thermal or chemical structure and is not subject to melting processes, which usually occur in the shallower region of the upper mantle [5]. According to Frost [6] the transition zone is characterized by some seismic discontinuities [7, 8] that should be related to mineral phase transformations. The phase assemblage of the transition zone mainly consists of ringwoodite, wadsleyite and majoritic garnet. Fig. 5.2 is a representation of the mineral volume-fraction in this layer (modified

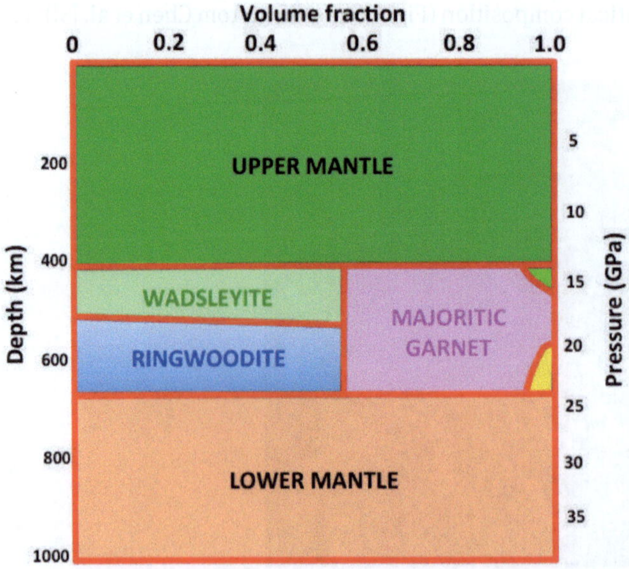

Fig. 5.2: Volume fraction of the transition zone between 410 and 660 km showing the three main phases ringwoodite, wadsleyite and, majoritic garnet (modified from [5]). In addition to the main phases diopside (light green) and in yellow $CaSiO_3$ (likely with walstromite-like structure) are displayed. Above and below the transition zone the upper mantle (dark green) and the lower mantle (light orange) are shown.

from Frost [5]). The diagram shows that the transition zone is constituted of about 41 % majoritic garnet, about 34 % ringwoodite, 23 % wadsleyite and the remaining 2 % are $CaSiO_3$ and $CaMgSi_2O_6$ diopside. In Fig. 5.2 the green area represents about 60 % olivine, which transforms to wadsleyite at a depth of about 410 km, which in turn transforms to ringwoodite at about 525 km [5]. Finally, ringwoodite breaks down to ferropericlase (Mg,Fe)O plus bridgmanite $MgSiO_3$ (perovskite-type structure). The reader is referred to chapter 4 for a more detailed discussion on structure and sharpness of the transition zone and thus the actual extension of the ringwoodite stability field.

5.2 Crystal structure and Mg/Fe substitution

For a better distinction from its lower-pressure polymorphs wadsleyite and common olivine, ringwoodite is also known as γ-olivine. Wadsleyite and olivine are called [9] β-olivine and α-olivine, with *Imma* and *Pbnm* symmetry (both orthorhombic), respectively. Thus, the two former modifications are symmetrically distinct from cubic γ-olivine of the $Fd\bar{3}m$ space group. Hereafter we will use the IMA mineral name ringwoodite, olivine (group name), and wadsleyite.

Ringwoodite has the chemical formula $(Mg,Fe)_2SiO_4$. The Fe dominant member was recently discovered in the Martian meteorite Tissint and named ahrensite ([10], IMA 2013-028, in honor of T.J. Ahrens (1936–2010), geophysicist at California Institute of Technology).

The ringwoodite-ahrensite solid solution shows complete miscibility as demonstrated by synthetic samples, e.g. [11–13]. Optical and physical properties for the two end-members are reported in Tab. 5.1. Ringwoodite and ahrensite crystallize in the spinel structure-type with space group $Fd\bar{3}m$. For comparison, the crystal structures of ringwoodite, olivine and wadsleyite are shown in Fig. 5.3.

Tab. 5.1: Physical and optical properties for ringwoodite and ahrensite.

	Ringwoodite	**Ahrensite**
Refractive index *n*	1.768	1.895*
Relief	High	High
Colour	Blue, deep blue, purple	Black
Calculated density	3.56 g/cm^3	4.98 g/cm^3
Hardness (Mohs)	7–7.5	

* estimated from the density differences between forsterite-fayalite and ringwoodite-ahrensite

Ringwoodite and olivine group minerals are both classified as orthosilicates. In these minerals, SiO_4 tetrahedra are isolated and not connected to other tetrahedra. In contrast, the crystal structure of wadsleyite has one oxygen (O2) connecting two tetrahe-

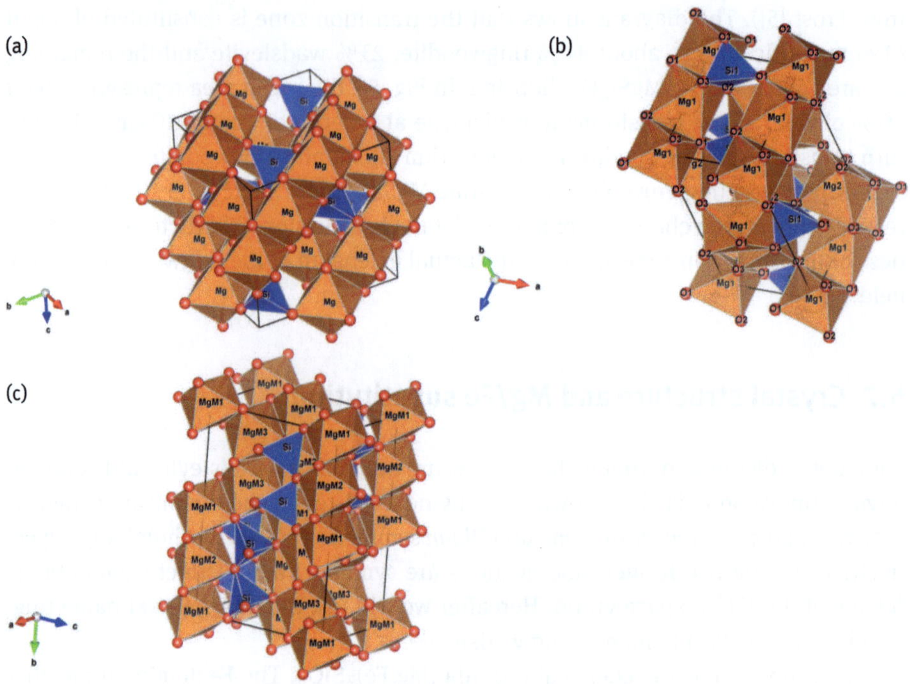

Fig. 5.3: Crystal structure of ringwoodite (a), forsterite (b) and wadsleyite (c). The structural data were from [11], [15] and [16], respectively. The orange octahedra are always relative to Mg, whereas the deep violet tetrahedra refer to Si. The red spheres correspond to oxygen atoms. The thin solid lines show the unit cell for each polymorph. The structures were drawn using Vesta software.

dra forming an Si_2O_7 group and is classified as sorosilicate. Ringwoodite has only one regular octahedron (with all Mg-O distances having the same length). Olivine has two (M1 and M2) more irregular octahedra, and wadsleyite has three symmetry independent M1, M2, and M3 octahedra (see Fig. 5.3).

Already in 1968 Kamb [14] explained on a structural basis, the reason why ringwoodite is the stable Mg_2SiO_4 form at high pressure. In case of olivine, wadsleyite, and ringwoodite the transformations are not driven by changes in the cation coordination number as observed for several other silicates. The olivine structure is based upon a non-ideal hexagonal-closest packed arrangement of oxygen atoms, with Si in tetrahedral coordination and Mg/Fe in octahedral coordination. In contrast, ringwoodite shows a cubic-closest packed arrangement of O atoms with Si and Mg/Fe having the same coordination number as in olivine. In general terms, this arrangement in olivine and ringwoodite should at first glance yield similar densities. This, however, is not the case. Actually, the density of ringwoodite is about 10 % higher than that of forsterite (olivine). We know that the higher density is the main reason for the stabilization of a polymorph to higher pressure, therefore, it is important to understand the structural reason for the higher density of end-member ringwoodite compared to end-member forsterite. The major reason for the density difference is the cation distribution in the second coordination sphere, considering also metal-metal (Si,Mg) distances [15]. The spinel structure exhibits six shared edges among adjacent octahedra whereas the olivine structure has on average (M1, M2) three shared octahedral edges and additionally three shared edges between MgO_6 octahedra and SiO_4 tetrahedra.

Only part of the density difference can be attributed to the average Mg-O distance, which is longer in olivine (about 2.120 Å, [15]) and significantly shorter in ringwoodite (2.066 Å, [11]). According to Kamb [14] the cause can be attributed to the shortening of shared polyhedral edges. The *c* axis in olivine is twice the Mg1-Mg1 distance (2.99 Å) across the shared octahedral edges whereas along the corresponding direction in ringwoodite the distance is only 2.85 Å contributing with a factor of 1.049 to the increased density (2.99 $Å_{olivine}/2.85_{ringwoodite}$ Å = 1.049). The length of the *a* axis in olivine is affected by the distortion of the SiO_4 tetrahedron. In particular, the distance from base to vertex of the tetrahedron in olivine is about 2.85 Å and only 2.7 Å in ringwoodite. Such a difference of 0.15 Å contributes to an elongation of the *a* axis in olivine by a factor of 1.033 (doubled distances: 4.70 $Å_{olivine}/4.55_{ringwoodite}$ Å = 1.033). If these two factors are combined with the difference in the average Mg-O distance (i.e. 2.12 $Å_{olivine}/2.066_{ringwoodite}$ Å = 1.027), an increased density $d = 3.59\ g/cm^3$ is estimated for ringwoodite relatively to the density of 3.23 g/cm^3 for forsterite. This simple calculation matches the experimental density of Mg ringwoodite.

The above reasoning can also be applied to the intermediate-pressure polymorph wadsleyite (d = 3.47 g/cm^3). Wadsleyite is considered a spinelloid and as such also based on a distorted cubic closest packing of O. M1 and M2 octahedra have six shared edges while M3 has seven. The average Mg-O distance in wadsleyite is 2.082 Å [16] while Mg-O in ringwoodite is 2.066 Å. This difference contributes to an increase of ringwoodite density by only 1 % (i.e. 2.082 $Å_{wadsleyite}/2.066_{ringwoodite}$ Å = 1.01). Along the *b* axis wadsleyite has a distance between Mg1 and Mg2 equal to 2.870 Å, which is slightly longer than that of ringwoodite along a corresponding direction. This difference only justifies a density increase by about 0.7 % (2.870 $Å_{wadsleyite}/2.850_{ringwoodite}$ Å = 1.007). However, along a direction close to [111] in wadsleyite the distance Mg1–Mg3 is 2.902 Å, while in ringwoodite the corresponding distance is only 2.850 Å (2.902 $Å_{olivine}/2.850_{ringwoodite}$ Å = 1.018). Thus, starting from the density of wadsleyite d = 3.47 g/cm^3, multiplied by 1.010 ×1.007 ×1.018 leads to d = 3.59 g/cm^3 for ringwoodite, again a perfect match with observation.

The structural parameters of ringwoodite and ahrensite are reported in Tab. 5.2. The available data on ringwoodite-ahrensite were collected [17] on synthetic samples as no natural crystals suitable for structural investigation have been found so far. This could be important to better understand whether in the Earth mantle the octahedral site in ringwoodite and ahrensite can be partly occupied by Si replacing Mg/Fe. Si in octahedral coordination, indeed, is a remarkable feature of some high-pressure minerals like majoritic garnet (i.e. [18, 19]) and stishovite (the high-pressure poly-

Tab. 5.2: Crystal-structure parameters for ringwoodite (Mg_2SiO_4) and ahrensite (Fe_2SiO_4).

Mineral	Cell edge *a* (Å)	Oxygen coordinate *u* (*x* = *y* = *z*)	Si coordinate (*x* = *y* = *z*)	Mg coordinate (*x* = *y* = *z*)	Si – O (Å)	V_{SiO4} (Å³)	Mg–O (Å)	V_{MgO6} (Å³)
Ringwoodite	8.0709(2)	0.2441(1)	0.125	0.5	1.6650(5)	2.369(2)	2.0664(3)	11.727(5)
				Fe coordinate (*x* = *y* = *z*)			Fe – O (Å)	V_{FeO6} (Å³)
Ahrensite	8.2312(2)	0.2420(4)	0.125	0.5	1.668(6)	2.382(10)	2.125(4)	12.732(34)

Notes: ringwoodite data were taken from [11], whereas the data for synthetic ahrensite are from [17]

morph of silica has Si only in octahedral coordination with average Si-O distance being 1.775 Å, see [20]). In particular, the case of majoritic garnet is of special importance as this phase is thought to be the most abundant phase of the transition zone: a typical natural majoritic garnet with respect to the ideal synthetic end-member, $Mg_3\,{}^{VI}(MgSi)Si_3O_{12}$, shows in the octahedral ^{VI}(MgSi) site about 0.3–0.4 Si atoms per formula unit (apfu) based on 12 O, Al close to 0.9–1.0 apfu, Fe^{2+} close to 0.5 apfu and some Ti^{4+} [19] However, Nakatsuka and coauthors [18] studying the structural change along the synthetic join $MgSiO_3 - Mg_3Al_2Si_3O_{12}$ found that for Si between 0.24 and 0.38 apfu the typical garnet symmetry $Ia\bar{3}d$ changes to tetragonal $I4_1/acd$. This was confirmed by diffraction data and supported by optical evidence. Independent of symmetry, in majoritic garnets Si occurs in both tetrahedral and octahedral coordination.

Ringwoodite synthetized at 20 GPa and 1400 °C may contain about 4 % of Si_{tot} at the octahedral site [21]. The presence of six-coordinated Si could be favored by high temperature [12] under these mantle-like pressure conditions. Surprisingly, for the Fe-richer samples Si does not show any tendency [11] to enter the octahedral site. It has not yet been investigated whether Si/Mg disorder is a peculiar characteristic of synthetic samples or whether it is also typical of natural samples. Hazen and coauthors [11] propose for their synthetic Mg end-member ringwoodite the crystal chemical formula $(Mg_{1.96}Si_{0.04})(Si_{0.96}Mg_{0.04})O_4$ and infer a double substitution ${}^{VI}Mg^{2+} + {}^{VI}Si^{4+} = {}^{IV}Mg^{2+} + {}^{IV}Si^{4+}$ to maintain charge balance.

5.3 The effect of water on the crystal structure of ringwoodite

In 1987 Smyth [22] suggested that hypothetically wadsleyite, the polymorph stable in the transition zone together with ringwoodite and majoritic garnet, could host a considerable amount of water and therefore possibly act as a significant mantle water reservoir. Although, there was actually no evidence of water (in terms of OH groups) in wadsleyite, Smyth [22] claimed that if this could be demonstrated then it could have substantial consequences for geophysical and geochemical models of the upper mantle. Just some years later, up to 3.1 wt.% H_2O in wadsleyite were experimentally demonstrated [9, 22, 23], confirming the hypothesis [22] about the capacity of such phases to host substantial amounts of water. At the same time, experimental evidence [9] demonstrated water solubility up to 2.7 wt.% H_2O in terms of OH groups in ringwoodite.

In the decade from 1985–1995 several experimental studies dealing with the synthesis of dense Mg-silicates focused on water storage under mantle conditions [24]. However, among all hydrous Mg-silicates only wadsleyite and ringwoodite could have a significant impact on the mantle geophysics and geochemistry and thus most projects focused on these two phases.

In the first study dealing with the influence of water on the ringwoodite structure, Smyth and coauthors [25] investigated seven samples of hydrous ringwoodite with compositions between Fo_{100} and Fo_{89} and H_2O contents between 0.2 and 1.1 wt.% [a typical composition of their sample was $Mg_{1.633}Fe^{2+}_{0.231}Fe^{3+}_{0.026}Si_{0.999}(H_{0.174})O_4$]. The samples, synthesized from 18 to 22 GPa and 1400 to 1500 °C, were analyzed by single-crystal X-ray diffraction, Fourier-transform infrared (FTIR) spectroscopy, Mössbauer spectroscopy, wave length dispersive spectroscopy (WDS) using an electron microprobe, and transmission electron-microscopy (TEM). It could be shown [26] that the principal hydration mechanism involves octahedral cation vacancies with the occupancy of the octahedral site appearing to decrease systematically with H content.

Results of Mössbauer studies indicated that Fe^{3+} is negligible (i.e. maximum content 0.026 apfu) and thus does not affect the crystal structure. The influence of Fe^{2+} on the structure of ringwoodite has already been discussed in the previous section. Let us consider the H-richest ringwoodite sample [25] SZ0104 with the chemical formula $Mg_{1.633}Fe^{2+}_{0.231}Fe^{3+}_{0.026}Si_{0.999}(H_{0.174})O_4$, which contains about 1.1 wt.% H_2O.

For anhydrous ringwoodite, complete replacement of Mg by Fe^{2+} (i.e. ahrensite end-member) causes an increase of the octahedral M-O (M = Mg/Fe) bond lengths by about 0.0006(3) Å/(0.01 molar fraction). Therefore, an increase in M-O = 2.0736 Å should be expected due to the Fe content in SZ0104, which is 0.129 in molar fraction. Instead the M-O bond length for hydrous ringwoodite SZ0104 is considerably longer with M-O = 2.0809(6) Å. The same logic applies to the unit-cell volume, which for a ringwoodite with starting composition $Mg_{1.742}Fe_{0.248}$, like SZ0104, should be 529.85(10) $Å^3$; instead the observed volume is 532.49(7) $Å^3$. Thus, in ringwoodite containing 1.1 wt.% H_2O the M-O distance is increased by 0.007 Å and the unit-cell volume by 2.6 $Å^3$.

More recently [24], Fe-free hydrous ringwoodite hosting 2.5 wt.% H_2O (measured by secondary ion mass spectroscopy) revealed a similar effect of H_2O on its crystal structure: the Mg-O distance increased by 0.006 Å and the unit-cell volume by 2.09 $Å^3$, relative to the prediction for the anhydrous composition. Panero et al. [26], studying similar samples at different pressure and temperature conditions, found different substitution mechanisms for OH groups in the ringwoodite structure. Both the Mg octahedra and Si tetrahedra may host H by a typical hydrogarnet substitution [26]. An increase in the unit-cell volume of 2.09 $Å^3$ due to 2.5 wt.% H_2O in ringwoodite corresponds to an expansion caused by a temperature increase of 140 °C.

5.4 Ringwoodite stability field

The Mg_2SiO_4-Fe_2SiO_4 system was studied at 1400 °C as a function of pressure and Fe/(Fe+Mg) ratio [6]. For the Mg end-member olivine transforms to wadsleyite slightly above 14 GPa and wadsleyite transforms to ringwoodite at about 20 GPa (Fig. 5.4). With increasing Fe the average transformation pressure decreases and at the same time the transformations occur within a significantly broader pressure interval. In general, it is observed (Fig. 5.4) that at 1400 °C the minimum pressure at which ringwoodite can exist reduces from 20 GPa for the pure Mg end-member to only about 6 GPa for the pure Fe end-member.

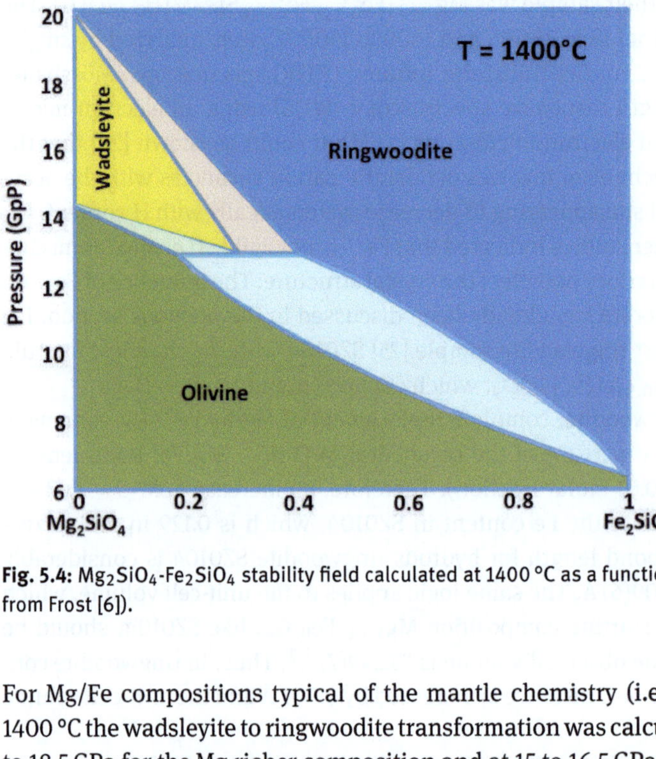

Fig. 5.4: Mg_2SiO_4-Fe_2SiO_4 stability field calculated at 1400 °C as a function of pressure (modified from Frost [6]).

For Mg/Fe compositions typical of the mantle chemistry (i.e. $(Mg_{0.9}Fe_{0.1})_2SiO_4$) at 1400 °C the wadsleyite to ringwoodite transformation was calculated [6] to occur at 17.5 to 18.5 GPa for the Mg-richer composition and at 15 to 16.5 GPa for the Mg-poorer composition. H_2O and Fe^{3+} are predicted to have little effect on the wadsleyite-ringwoodite transformation as they have similar solubility in both wadsleyite and ringwoodite [6].

Thus, ringwoodite and wadsleyite can coexist between 16.5 and 17.5 GPa (about 495 and 525 km depth) at 1285 °C and between about 17.7 and 18.7 GPa (about 531 and 560 km depth) at 1470 °C. The maximum pressure at which ringwoodite can still exist (where wadsleyite is no longer stable) is defined by the ringwoodite breakdown to ferropericlase + bridgmanite, which is widely agreed to occur at an average depth of 660 km with an uncertainty of about 30–50 km (i.e. [5]).

5.5 Thermo-elastic properties of ringwoodite

5.5.1 Bulk modulus

The bulk modulus, defined as $K = -V\partial P/\partial V$, is a key thermodynamic parameter being the inverse of compressibility, which indicates the volume change as a function of pressure. The bulk modulus is a central property that provides information about elasticity of Earth's materials [27]. In order to determine the bulk modulus of ringwoodite we should consider only methods suitable for studying "small samples" in the 0.01–0.1 mm range. Among these techniques, the most widely used are:

(1) in-situ high-pressure X-ray diffraction, which allows under static compression conditions, determination of K_T by studying the unit-cell volume at different pressures and then fitting the pressure-volume data to some equation of state (see [28], EoSFIT7.c).
(2) Brillouin scattering provides the adiabatic bulk modulus K_S and is based on the relationship between the acoustic elastic waves and the elastic moduli and density.
(3) Ultrasonic interferometry is based on wave velocity measurement; the seismic wave velocities Vp and Vs are related to K_S, to the shear modulus G, and to the density of a material.

Finally, in addition to experimental studies there are several computational approaches that should be considered; however, a detailed discussion of these is beyond the scope of this review.

In order to compare the data obtained using the three techniques listed above, an expression relating K_S and K_T is required: $K_S = K_T(1 + \alpha\gamma T)$, with α = thermal expansion, γ = Grüneisen parameter and T = temperature. At low temperature the difference between K_S and K_T is not larger than about 1 %, which is of the same magnitude as the experimental uncertainty. However at high temperature such difference becomes significant and the conversion factor $(1 + \alpha\gamma T)$ must be taken into account. For the purpose of a simplified comparison [29] at $T = 298$ the conversion factor is 1.099 for ringwoodite.

Surprisingly, more research has been performed on the effect of water on the bulk modulus of ringwoodite than focusing on the same mineral under dry conditions. This is likely due to the discovery that ringwoodite can host significant amounts of water. In the following paragraph the bulk modulus (and thermal expansion) of dry ringwoodite is analyzed and then the analysis is extended to data from hydrous ringwoodite.

In Tab. 5.3 the values of adiabatic bulk modulus (K_T is converted in K_S by using the conversion factor of 1.099) and its first pressure derivative (where available) are reported for dry ringwoodite. In detail, data for Mg and Fe end-members and intermediate compositions are separated. Analyzing Tab. 5.3 Mg ringwoodite shows general agreement among different studies with an average value of K_S = 191(11) GPa. The

Tab. 5.3: Adiabatic bulk modulus and its first pressure derivative for the ringwoodite-ahrensite solid solution. The data of bulk modulus obtained by X-ray diffraction were converted in adiabatic bulk modulus for purpose of comparison among different techniques.

Composition	*Ks*	*K′*	Technique	Reference
Mg_2SiO_4	215	4 (fixed)	X-ray diffraction	[43]
Mg_2SiO_4	184	–	Brillouin	[44]
Mg_2SiO_4	186	4.8 (fixed)	X-ray diffraction	[11]
Mg_2SiO_4	184	4.2	X-ray diffraction	[44]
Mg_2SiO_4	199	4.19	Computational	[45]
Mg_2SiO_4	184	–	Brillouin	[46]
Mg_2SiO_4	185	–	Brillouin	[47]
Mg_2SiO_4	206	4.51	Computational	[48]
Mg_2SiO_4	185	4.5	Ultrasonic	[49]
Mg_2SiO_4	184	4.4	Ultrasonic	[50]
Mg_2SiO_4	185	4.2	Ultrasonic	[51]
$Mg_{1.68}Fe_{0.32}SiO_4$	176	4 (fixed)	X-ray diffraction	[52]
$Mg_{1.84}Fe_{0.16}SiO_4$	188	4.1	Brillouin	[53]
$Mg_{1.8}Fe_{0.2}SiO_4$	192	4.7	X-ray diffraction	[54]
Fe_2SiO_4	209	4.8 (fixed)	X-ray diffraction	[11]
Fe_2SiO_4	220	–	Brillouin	[46]
Fe_2SiO_4	200	4.4	Ultrasonic	[50]
Fe_2SiO_4	203	4.4	Ultrasonic	[51]
Fe_2SiO_4	189	5.5	X-ray diffraction	[13]
Fe_2SiO_4	203	3.7	X-ray diffraction	[55]

first pressure derivative K' on average is 4.35. For intermediate Mg-dominant compositions a similar behavior is observed as for the Mg-end member. For ahrensite (Fe_2SiO_4) a large data scatter with values ranging from 187 to 220 GPa is found. The average value of the adiabatic bulk modulus K_S is 204(10) GPa. The first pressure derivative K' is on average 4.56, in good agreement with the Mg end-member.

Fitting all data in Tab. 5.3 the following linear equation is obtained:

$$K_S\,(\text{GPa}) = 203.73 - 0.1447 \times (\%\ Mg_2SiO_4) \tag{5.1}$$

The first pressure derivative K' shows a general agreement with an average value of 4.4(4). The evolution of K_S as a function of Mg_2SiO_4 end-member is displayed in Fig. 5.5.

5.5.2 The effect of water on the bulk modulus of ringwoodite

As ringwoodite can host a considerable amount of structural water in terms of OH groups and this significantly affects its crystal structure, it is quite reasonable to assume that water can also affect the thermodynamic properties.

In Tab. 5.4 all available data on hydrous ringwoodite with Fe up to about 50 % and water content ranging from 0.40 to 2.80 H_2O wt.% are reported. In a first step Fe-free

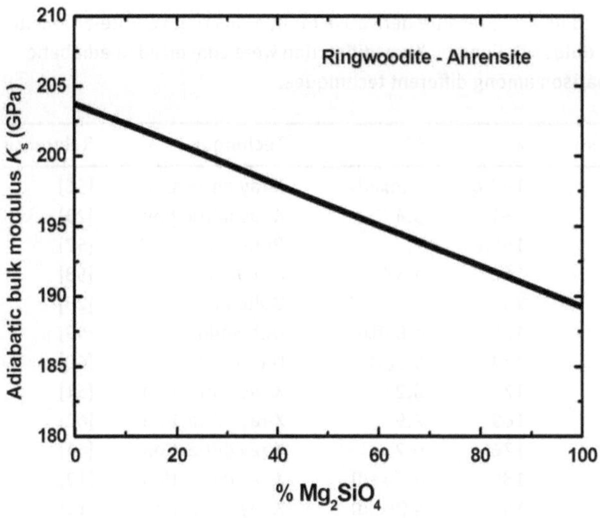

Fig. 5.5: Evolution of the adiabatic bulk modulus, K_S, as a function of the Mg/Fe substitution along the ringwoodite-ahrensite solid solution. The solid line was obtained by using the data in Tab. 5.1 and the relationship (5.1) reported in the text.

ringwoodite with water content between 2.20 and 2.80 % (Tab. 5.2) is considered. The bulk modulus ranges from 165.8 to 149.4 GPa, with an average value of 159(7) GPa and an average $K' = 4.9(4)$.

For Fe-richer samples with 1 % H_2O, the bulk modulus increases to about 176–177 GPa and for samples with 0.4–0.7 % H_2O the bulk modulus reaches the average value of Fe-free anhydrous ringwoodite (Mg_2SiO_4). In order to quantify the effect of water on the bulk modulus for Fe-free ringwoodite the following equation was obtained:

$$K_S\,(\text{GPa}) = 189.8 - 12.607 \times (H_2O\ \text{wt.\% in}\ Mg_2SiO_4) \tag{5.2}$$

using data in Tab. 5.4 for hydrous ringwoodite and the average bulk modulus of equation (5.1) for a Fe-free ringwoodite. In Fig. 5.6 the variation of K_S for pure Mg_2SiO_4 with the water content is shown.

However, natural ringwoodite is always Fe-bearing requiring knowledge of the effect of H_2O on a Fe-bearing variety. This could be attempted by combining equations (5.1) and (5.2). For equation (5.1) the presence of Fe increases the bulk modulus, whereas for equation (5.2) the presence of water decreases K_S.

Thus, equation (5.3) provides an empirical way of obtaining the adiabatic bulk modulus of ringwoodite as a function of Fe and H_2O content. This equation (5.3) reproduces the data in Tab. 5.4 within an uncertainty of about ±4 GPa if all uncertainties are propagated:

$$K_S\,(\text{GPa}) = [203.73 - 0.1447 \times (\%\ Mg_2SiO_4) - 12.607 \times (H_2O\ \text{wt.\% in}\ Mg_2SiO_4)] \tag{5.3}$$

Tab. 5.4: Adiabatic bulk modulus and its first pressure derivative for hydrous ringwoodite-ahrensite solid solution. The data of bulk modulus obtained by X-ray diffraction were converted in adiabatic bulk modulus for purpose of comparison among different techniques.

Composition	Water content	K_s	K'	Technique	Reference
Mg_2SiO_4	2.80 %	149.4	5 (fixed)	X-ray diffraction	[56]
Mg_2SiO_4	2.50 %	161	5.4	X-ray diffraction	[23]
Mg_2SiO_4	2.30 %	165.8	–	Brillouin	[57]
Mg_2SiO_4	2.30 %	166	4.37	Brillouin	[58]
Mg_2SiO_4	2.20 %	155	–	Brillouin	[23]
$Mg_{1.76}Fe_{0.24}SiO_4$	1.00 %	176	4.8 (fixed)	Ultrasound	[59]
$Mg_{1.76}Fe_{0.24}SiO_4$	1.00 %	177	5.3(4)	Ultrasound	[60]
$Mg_{1.76}Fe_{0.24}SiO_4$	1.00 %	177	6.2	X-ray diffraction	[61]
$Mg_{1.78}Fe_{0.22}SiO_4$	0.93 %	169	7.9	X-ray diffraction	[62]
$Mg_{1.90}Fe_{0.10}SiO_4$	0.79 %	176	6.2	X-ray diffraction	[61]
$Mg_{0.98}Fe_{1.02}SiO_4$	0.70 %	188	4 (fixed)	X-ray diffraction	[12]
$Mg_{1.22}Fe_{0.78}SiO_4$	0.40 %	186	4 (fixed)	X-ray diffraction	[12]

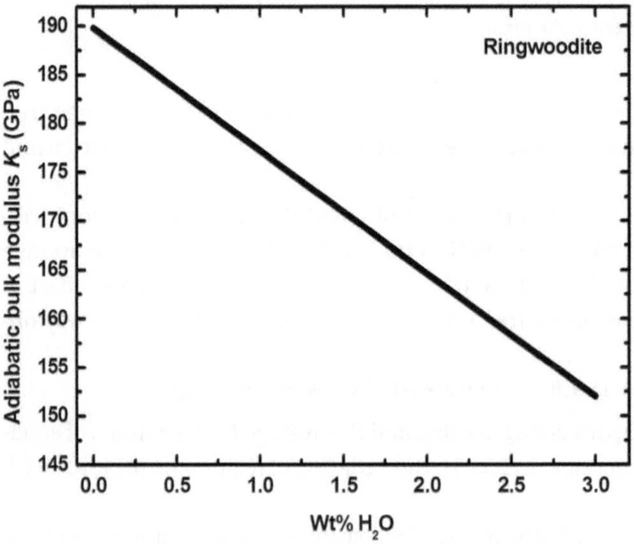

Fig. 5.6: Evolution of the adiabatic bulk modulus, K_S, as a function of the water content in Mg_2SiO_4 ringwoodite. The solid line was obtained by using the data in Tab. 5.2 and the relationship (5.2) reported in the text.

The calculation of (5.3) should not be affected by the first pressure derivative of K_S as the average value for K' for anhydrous ringwoodite is 4.4(4) and 4.9(7) for hydrous ringwoodite. A simple average of these values and their standard deviations (K' = 4.7(8)) should yield a reliable result for any ringwoodite composition.

5.5.3 Thermal expansion of anhydrous and hydrous ringwoodite

Thermal expansion, together with compressibility, represents one of the most important thermodynamic parameters. It could be simply defined as the volume variation as a function of temperature at a constant pressure being $\alpha(T) = V^{-1}(\partial V/\partial T)$ (with the only constraint for such expression being $\alpha(T) = (\partial\alpha/\partial T) = 0$ at absolute zero, e.g. [28]). Different thermal expansion equations might be used: the Berman equation, the Fei equation, the Salje equation, the modified Holland and Powell equation or the Kroll form of the Holland and Powell equation as described in detail in [28]. Literature data are applied in this review to determine the volume thermal-expansion coefficient at 298 K, α_{V0} for anhydrous ringwoodite. Only one anhydrous intermediate composition between ringwoodite and ahrensite has been taken into account.

The thermal expansion data for "dry" and hydrous ringwoodite are reported in Tab. 5.5. The data for anhydrous Mg_2SiO_4 show in general significant scatter but the average value is $\alpha_{V0} = 2.3(5) \times 10^{-5}$ K^{-1}. The intermediate composition, $(Mg_{0.91}Fe_{0.09})_2SiO_4$, shows a similar volume thermal-expansion and also anhydrous Fe_2SiO_4 has an average value of $\alpha_{V0} = 2.3(2) \times 10^{-5}$ K^{-1}. Apparently, there is no change in thermal expansion along the ringwoodite-ahrensite solid-solution series.

Tab. 5.5: Volume thermal expansion data for anhydrous and hydrous ringwoodite-ahrensite solid solution.

Composition	Water content	α_{V0} ($\times 10^{-6}$/K)	Technique	Reference
Mg_2SiO_4	0	1.90	X-ray diffraction	[63]
Mg_2SiO_4	0	2.37	Calorimetry	[64]
Mg_2SiO_4	0	1.77	X-ray diffraction	[47]
Mg_2SiO_4	0	2.70	HPVS*	[65]
Mg_2SiO_4	0	3.07	X-ray diffraction	[24]
Mg_2SiO_4	0	2.57	X-ray diffraction	[66]
Mg_2SiO_4	0	1.97	Computational	[67]
Mg_2SiO_4	0	2.74	Computational	[68]
Mg_2SiO_4	0	1.75	Computational	[69]
$Mg_{1.82}Fe_{0.18}SiO_4$	0	2.50	X-ray diffraction	[70]
Fe_2SiO_4	0	2.30	X-ray diffraction	[71]
Fe_2SiO_4	0	2.46	Thermochemical	[72]
Fe_2SiO_4	0	2.62	X-ray diffraction	[73]
Fe_2SiO_4	0	2.15	Computational	[74]
Fe_2SiO_4	0	2.13	Computational	[75]
Mg_2SiO_4	2.60 %	2.73	X-ray diffraction	[23]
Mg_2SiO_4	2.50 %	2.90	X-ray diffraction	[24]

*High-pressure vibrational spectroscopy

Few results [30] are not covered by Tab. 5.5. The excluded data were affected by "disequilibrium irreversible expansion starting already at 606 K" [30]. For water contents between 2.50 and 2.60 wt.% H_2O the average thermal expansion coefficient is $\alpha_{V0} = 2.8(1) \times 10^{-5}\,K^{-1}$. Although only two data sets are used for the present comparison, it seems that water increases the thermal expansion of ringwoodite by about 17–18 %. Combining the data in Tab. 5.5 the following equation is derived:

$$\alpha_{V0} = 2.316 \times 10^{-5}\,K^{-1} + 0.195 \times (\text{wt.\% } H_2O \text{ in } Mg_2SiO_4). \tag{5.4}$$

Equation (5.4) thus takes into account the effect of Fe and water on the thermal expansion coefficient of ringwoodite.

5.5.4 Thermo-elastic properties of ringwoodite: implications for Earth Sciences

How could the change of the bulk modulus and of the thermal expansion as a function of composition and water content be used in Earth Sciences obtaining geophysical information? It was recently demonstrated that the new approach of so-called "elastic geobarometry" can be applied to any mineral inclusion entrapped in a mineral host. Such an approach, developed by Angel et al. [31], is mainly based on the difference of thermo-elastic properties between host and inclusion and on the residual pressure to which the inclusion is still exposed when the host is at atmospheric pressure. A very good example is represented by diamond and its mineral inclusions. Due to the very high bulk modulus and very low thermal expansion coefficient, diamond shows a much smaller expansion when it reaches the surface from the mantle. In contrast, the inclusions in diamond tend to expand much more as their thermo-elastic properties are strongly different from those of diamond. In general, diamond has at least a 2–3 times larger bulk modulus than common silicates found in it as inclusions and a thermal expansion, which is at least 3 times smaller [31]. This means that, for those diamond-inclusion pairs not showing any cracks at atmospheric pressure, the inclusion is always under some pressure (i.e. 0.4–0.5 GPa, see [15]), which is called "residual or remnant or internal pressure".

If such a geobarometric method is applied to the diamond-ringwoodite pair important information on the possible depth of entrapment can be obtained. The software EosFit7c [28] was used to perform the calculation of the elastic properties for a pure Mg anhydrous ringwoodite, a Fe-rich ringwoodite, and a very hydrous ringwoodite like the one found in the Brazilian diamond by Pearson et al. [32].

A ringwoodite host within diamond, not surrounded by cracks, is assumed to have a possible internal pressure of 5 GPa. For a temperature of 2000 °C in the transition zone, a pressure of formation (or entrapment) for the diamond-ringwoodite pair can be calculated that is equal to 22.7 GPa for a pure anhydrous Mg ringwoodite. Adding some Fe (20 %) to the anhydrous ringwoodite such calculation provides a pressure of 22.9 GPa, so the difference is not significant as the bulk modulus and the thermal expansion as a function of the Mg/Fe substitution do not show large variations (at least for limited substitutions). If instead, the same calculation is performed for a hydrous ringwoodite, with a water concentration (ca. 1.4 wt.%) similar to the one found in a diamond inclusion [32], then at 2000 °C the pressure of formation is reduced to 17.2 GPa. These results are of high importance because it is known that the pressure of formation obtained above for anhydrous ringwoodite (regardless the Fe content at least up to 20 % in molar fraction of Fe_2SiO_4) are too high and fall in the ferropericlase + bridgmanite stability field. In contrast, the pressure of formation obtained for hydrous ringwoodite leads to the transition zone at about 515–516 km depth, which nearly overlaps with the wadsleyite – ringwoodite phase boundary. Thus, this is proof that the presence of water decreases the pressure of the ringwoodite stability field. However, the true values are strongly dependent on the quality of the thermo-elastic parameters. To conclude, we hope that in future it will be possible to measure the unit-cell parameters and the crystal structure of a natural ringwoodite still trapped in a diamond in order to obtain reliable geobarometric data.

5.6 Ascent of diamond-bearing kimberlite magma

The discovery of ringwoodite in diamond not only provided a new scenario relatively to the real amount of water stored in Earth but at the same time it could provide new constraints on the debate about the ascent velocity of diamond-bearing kimberlite magmas. This topic has been strongly discussed in the literature and still doubts remain. The literature data suggest an ascent velocity from some meters per second to tens of meters per second [33].

One of the most recent contributions to this issue originates from Baruah et al. [34]. These authors tackled the "rapid ascent problem" by a multi-directional approach: they took into account (a) the kinetics of the diamond-graphite transformation, (b) the settling velocity of diamond phenocrysts in magmas, (c) the formation of ruptures during high-speed magma ascent. Such an approach is reasonable, as none of the above points must be neglected to provide a reliable value of ascent velocity.

In detail, Baruah et al. [34] measured the degree of graphitization of diamond as a function of the ascent velocity in a synthetic kimberlitic diamond-bearing magma (also varying the starting compositions). Even if the authors cannot provide a "single number" for their final results they propose an ascent velocity higher than 10 m/s. It must be remarked that even for this ascent velocity they find some degree of graphitization on very small diamonds. Unfortunately, the authors [34] did not provide measurements for velocities higher than 10 m/s. Averaging their measured data, a reasonable velocity for large diamond with no evidence of graphitization can be estimated to be around 20 m/s (or about 72 km/h). Octahedral lithospheric diamond, up to several mm large, with no traces of graphitization is quite common (see [35]). For such diamonds the ascent time may be estimated to last a couple of hours assuming a possible depth of formation at around 150 km [15].

If for "super-deep" diamonds similar ascent velocities are assumed, such diamond travels from the transition zone, about 550 km, to the surface in less than 8 hours. It is known from ringwoodite found in a Brazilian diamond that its host diamond crystallized at least at a depth of 525 km. However, it remains unknown whether the diamond reached the Earth's surface in one single "trip" from 525 km or this long journey had multiple stages, characterized by a first very fast ascent reaching some depths under the cratonic area at which the temperature was too low to allow back-transformation of ringwoodite to olivine. Unfortunately, there are no kinetic data that could reveal how fast the back-transformation of ringwoodite to olivine is, and at the same time it needs to be considered that such a transformation should occur inside diamond. Thus, for a meaningful study of such kinetic processes it will be necessary to obtain large diamond crystals containing ringwoodite and at present such experiments appears to be highly unlikely.

5.7 Ringwoodite and the Earth's storage water capacity

The Earth's water storage capacity is one of the most debated scientific issues in Earth Sciences. For decades scientist have been trying to find out how much water is stored in our planet. Several investigations addressed this topic and one of the recent reviews [36] indicates a total mass of water equal to 3.04×10^{24} g. An even more recent review [37] recalculates the mass of water at 5.50×10^{24} g. This difference is primarily due to the crucial discovery [32] of a crystal of hydrous ringwoodite enclosed within a "super-deep" diamond from Brazil characterized by a water concentration of about 1.4 wt.% (Fig. 5.7). A recent study [38] confirmed this water concentration by using a new IR calibration and reported a concentration of 1.43(27) wt.% H_2O for the ringwoodite inclusion [32]. This water concentration represents about 50 % of the maximum water solubility in ringwoodite and is different from previous assumptions about natural ringwoodite compositions. Furthermore, if ringwoodite is hydrous there are no reasons to think that wadsleyite is not. Even if these two minerals in the transition zone could host half of their maximum water solubility, as found in laboratory, such a finding could change the complete scenario of the water cycle on our planet. This discovery would also explain why, compared to the amount water determined for CI carbonaceous chondritic meteorites, which is about 10 % ([39] and references therein), the Earth appeared so dry. Apparently the difference between the water content on the Earth and that in CI chondrites is much smaller than that previously estimated and this could be due to ringwoodite and wadsleyite in the transition zone.

Some authors actually suggested that ringwoodite found in diamond may only represent a locally water-rich enrichment and be unrepresentative of the real water concentration of the transition zone (e.g. [40]). Although this explanation cannot be

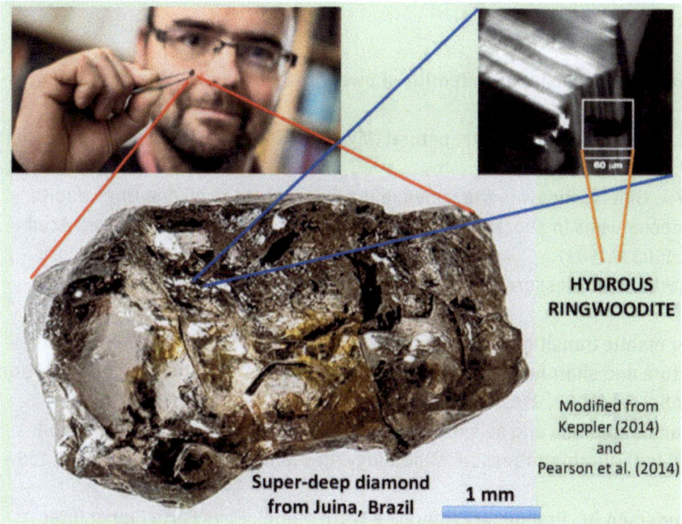

Fig. 5.7: Dr. Graham Pearson (top left) holding the diamond from Juina in which for the first time hydrous ringwoodite was found in the Earth [32]. On bottom left a magnification of the diamond and on top right a magnification of the tiny hydrous ringwoodite grain. The images are modified from [32] and [40]. The picture of the diamond is from Richard Siemens (University of Alberta).

ruled out, a recent study on the transformation of hydrous ringwoodite to $MgSiO_3$ bridgmanite plus (Mg,Fe)O (occurring at 660 km depth) documented the formation of intergranular melt and this might be consistent with large hydrated regions in the transition zone [41]. Moreover, a strongly hydrated transition zone has been proposed by geophysicists (i.e. [42]) and although the level of hydration of the transition zone has not been proven conclusively, thanks to ringwoodite any future geophysical model of this region and of the entire Earth must take into account much more water than previously thought.

Acknowledgment: The research is funded by the European Research Council to FN (ERC Starting Grant INDIMEDEA, grant agreement no. 307322). Antony Burnham from Research School of Earth Sciences of the Australian National University is thanked for his complete review of the work. Matteo Alvaro is strongly thanked for his contribution in the pressure of formation calculation on the diamond-ringwoodite pair and for several scientific discussions.

Bibliography

[1] Spencer LJ. The Tenham (Queensland) meteoritic shower of 1879. Mineral Mag 1937, 24, 437–457.

[2] Binns RA, Davis RJ, Reed SJB. Ringwoodite, natural $(Mg,Fe)_2SiO_4$ spinel in the Tenham meteorite. Nature 1969, 221, 943–944.

[3] Chen M, El Goresy A, Gillet P. Ringwoodite lamellae in olivine: clues to olivine-ringwoodite phase transition mechanisms in shocked meteorites and subducting slabs. Proc Natl Acad Sci USA 2004, 42, 15033–15037.

[4] Ringwood AE, Major A. High pressure transformations of spinels (I). Earth Planet Sci Lett 1966, 5, 242–250.

[5] Frost DJ. The upper mantle transition zone. Elements 2008, 4, 171–176.

[6] Frost DJ. The structure and sharpness of $(Mg,Fe)_2SiO_4$ phase transformations in the transition zone. Earth Planet Sci Lett 2003, 216, 313–328.

[7] Agee CB. Phase transformations and seismic structure in the upper mantle and transition zone. In: Hemley RJ (ed) Ultrahigh-Pressure Mineralogy. Rev Mineral Geochem 1998, 37, 165–203.

[8] Dziewonski AM, Anderson DL. Preliminary reference Earth model. Phys Earth Planet Inter 1981, 25, 297–356.

[9] Kohlstedt DL, Keppler H, Rubie DC. Solubility of water in the α, β, and γ phases of $(Mg,Fe)_2SiO_4$. Contrib Mineral Petrol 1996, 123, 345–357.

[10] Ma C, Tschauner O, Liu Y, Sinogeikin S. Ahrensite, IMA 2013-028. CNMNC Newsletter No. 16, August 2013, page 2707; Mineral Mag 2013, 77, 2695–2709.

[11] Hazen RM, Downs RT, Finger LW, Ko J. Crystal chemistry of ferromagnesian silicate spinels: evidence for Mg-Si disorder. Am Mineral 1993, 78, 1320–1323.

[12] Ganskow G, Boffa Ballaran T, Langenhorst F. Effect of iron on the compressibility of hydrous ringwoodite. Am Mineral 2010, 95, 747–753.

[13] Nestola F, Boffa Ballaran T, Koch-Mueller M, Balic-Zunic T, Taran M, Olsen L, Princivalle F, Secco L, Lundegaard L. New accurate compression data for γ-Fe_2SiO_4. Phys Earth Planet Inter 2010, 183, 421–425.

[14] Kamb B. Structural basis of the olivine-spinel stability relation. Am Mineral 1968, 53, 1439–1455.

[15] Nestola F, Nimis P, Ziberna L, Longo M, Marzoli A, Harris JW, Manghnani MH, Fedortchouk Y. First crystal-structure determination of olivine in diamond: composition and implications for provenance in the Earth's mantle. Earth Planet Sci Lett 2011, 305, 249–255.

[16] Holl CM, Smyth JR, Jacobsen SD, Frost DJ. Effects of hydration on the structure and compressibility of wadsleyite β-(Mg_2SiO_4). Am Mineral 2008, 93, 598–607.

[17] Nestola F, Balic-Zunic T, Koch-Mueller M, Secco L, Princivalle F, Parisi F, Dal Negro A. High-pressure crystal structure investigation of synthetic Fe_2SiO_4 spinel. Mineral Mag 2011, 75, 2649–2655.

[18] Nakatsuka A, Yoshiasa A, Yamanaka T, Ohtaka O, Katsura T, Ito E. Symmetry change of majorite solid-solution in the system $Mg_3Al_2Si_3O_{12}$ – $MgSiO_3$. Am Mineral 1999, 82, 1135–1143.

[19] Collerson KD, Hapugoda S, Kamber BS, Williams Q. Rocks from the mantle transition zone: majorite-bearing xenoliths from Malaita, Southwest Pacific. Science 2000, 288, 1215–1223.

[20] Yamanaka T, Fukuda T, Tsuchiya J. Bonding character of SiO_2 stishovite under high pressures up to 30 GPa. Phys Chem Miner 2002, 29, 633–641.

[21] Gasparik T. The role of volatiles in the translation zone. J Geophys Res 1993, 98, 4287–4299.

[22] Smyth JR. β-Mg_2SiO_4; a potential host for water in the mantle? Am Mineral 1987, 72, 1051–1055.

[23] Inoue T. Effects of water on melting phase relations and melt composition in the system Mg_2SiO_4-$MgSiO_3$-H_2O to 15 GPa. Phys Earth Planet Inter 1994, 85, 237–263.
[24] Ye Y, Brown DA, Smyth JR, Panero WR, Jacobsen SD, Chang YY, Townsend JP, Thomas SM, Hauri EH, Dera P, Frost DJ. Compressibility and thermal expansion of hydrous ringwoodite with 2.5(3) wt.% H_2O. Am Mineral 2012, 97, 573–582.
[25] Smyth J, Holl CM, Frost DJ, Jacobsen SD, Langenhorst F, McCammon C. Structural systematics of hydrous ringwoodite and water in Earth's interior. Am Mineral 2003, 88, 1402–1407.
[26] Panero WR, Smyth JR, Pigott JS, Liu Z, Frost DJ. Hydrous ringwoodite to 5 K and 35 GPa: Multiple hydrogen bonding sites resolved with FTIR spectroscopy. Am Mineral 2013, 98, 637–642.
[27] Angel RJ, Jackson JM, Reichmann HJ, Speziale S. Elasticity measurements on minerals: a review. Eur J Mineral 2009, 21, 525–550.
[28] Angel RJ, Gonzalez-Platas J, Alvaro M. EosFit7c and a Fortran module (library) for equation of state calculations. Z Kristallogr 2014, 229, 405–419.
[29] Li L, Weidner DJ, Brodholt J, Alfe D, Price GD. Elasticity of Mg_2SiO_4 ringwoodite at mantle conditions. Phys Earth Planet Inter 2006, 157, 181–187.
[30] Ye Y, Schwering RA, Smyth JR. Effects of hydration on thermal expansion of forsterite, wadsleyite, and ringwoodite at ambient pressure. Am Mineral 2009, 94, 899–904.
[31] Angel RJ, Alvaro M, Nestola F, Mazzucchelli ML. Diamond thermoelastic properties and implications for determining the pressure of formation of diamond-inclusion systems. Russ Geol Geophys 2015, 56, 211–220.
[32] Pearson DG, Brenker FE, Nestola F, McNeill J., Nasdala L, Hutchison MT, Matveev S, Mather K, Silversmit G, Schmitz S, Vekemans B, Vincze L. Hydrous mantle transition zone indicated by ringwoodite included within diamond. Nature 2014, 507, 221–224.
[33] Sparks RSJ, Baker L, Brown RJ, Field M, Schumacher J, Stripp G, Walters A. Dynamical constraints on kimberlite volcanism. J Volcanol Geotherm Res 2006, 15, 18–48.
[34] Baruah A, Gupta AK, Mandal N, Singh RN. Rapid ascent conditions of diamond-bearing kimberlitic magmas: Findings from high pressure–temperature experiments and finite element modeling. Tectonophysics 2013, 594, 13–26.
[35] Nestola F, Nimis P, Angel RJ, Milani S, Bruno M, Prencipe M, Harris JW. Olivine with diamond-imposed morphology included in diamonds. Syngenesis or protogenesis? Int Geol Rev 2014, 56, 1658–1667.
[36] Bodnar RJ, Azbej T, Becker SP, Cannatelli C, Fall A, Severs MJ. Whole Earth geohydrologic cycle, from the clouds to the core: The distribution of water in the dynamic Earth system, *in* Bickford, M.E., ed., The Web of Geological Sciences: Advances, Impacts, and Interactions: Spec Pap – Geol Soc Am 2013, 500, 431–461.
[37] Nestola F, Smyth JR. Diamonds and water in the deep Earth: a new scenario. Int Geol Rev 2015, in press. DOI: 10.1080/00206814.2015.1056758.
[38] Thomas SM, Jacobsen SD, Bina CR, Reichart P, Moser M, Hauri EH, Koch-Müller M, Smyth, JR, Dollinger G. Quantification of water in hydrous ringwoodite. Front Earth Sci 2015, doi: 10.3389/feart.2014.00038.
[39] Kerridge F. Carbon, hydrogen and nitrogen in carbonaceous chondrites: abundances and isotopic compositions in bulk samples. Geochim Cosmochim Acta 1985, 49, 1707–1714.
[40] Keppler H. Earth's deep water reservoir. Nature 2014, 507, 174–175.
[41] Schmandt B, Jacobsen SD, Becker TW, Zhenxian L, Dueker KG. Dehydration melting at the top of lower mantle. Science 2014, 344, 1265–1268.
[42] Bercovici D, Karato S. Whole-mantle convection and the transition-zone water filter. Nature 2003, 425, 39–44.
[43] Mizukami S, Ohtani A, Kawai N. High-pressure X-ray diffraction studies on β- and γ-Mg2SiO4. Phys Earth Planet Inter 1975, 10, 177–182.

[44] Meng Y, Weidner DJ, Gwanmesia GD, Liebermann RC, Vaughan MT, Wang Y, Leinenweber K, Pacalo RE, Yeganeh-Haeri A, Zhao Y. In Situ high P-T X-ray diffraction studies on three polymorphs (α, β, γ) of Mg_2SiO_4. J Geophys Res 1993, 98, 22199–22207.
[45] Kiefer B, Stixrude L, Wentzcovitch RM. Calculated elastic constants and anisotropy of Mg_2SiO_4 spinel at high pressure. Geophys Res Lett 1997, 24, 2841–2844.
[46] Sinogeikin SV, Katsura T, Bass JD. Sound velocities and elastic properties of Fe-bearing wadsleyite and ringwoodite. J Geophys Res 1998, 103, 20819–20825.
[47] Jackson JM, Sinogeikin SV, Bass JD. Sound velocities and elastic properties of γ-Mg_2SiO_4 to 873 K by Brillouin spectroscopy. Am Mineral 2000, 85, 296–303.
[48] Bukowinski MST, Downs JW. Structures, compressibilities and relative stabilities of the alpha, beta and gamma phases of Mg_2SiO_4 deduced from an electron-gas ionic Hamiltonian. Geophys J Int 2000, 143, 295–301.
[49] Li B. Compressional and shear wave velocities of ringwoodite γ-Mg_2SiO_4 to 12 GPa. Am Mineral 2003, 88, 1312–1317.
[50] Higo Y, Inoue T, Li BS, Irifune T, Liebermann RC. The effect of iron on the elastic properties of ringwoodite at high pressure. Phys Earth Planet Inter 2006, 159, 276–285.
[51] Liu Q, Liu W, Whitaker ML, Wang L, Li B. Compressional and shear wave velocities of Fe_2SiO_4 spinel at high pressure and high temperature. High Pressure Res 2008, 28, 405–413.
[52] Fei YW, Mao HK, Shu JF, Parthasarathy G, Bassett WA, Ko JD. Simultaneous high-P, high-T X-ray-diffraction study of beta-$(Mg,Fe)_2SiO_4$ to 26-GPa and 900-K. J Geophys Res 1992, 97, 4489–4495.
[53] Sinogeikin SV, Bass JD, Katsura T. Single-crystal elasticity of ringwoodite to high pressures and high temperatures: Implications for 520 km seismic discontinuity. Phys Earth Planet Inter 2003, 136, 41–66.
[54] Matsui M, Katsura T, Kuwata A, Hagiya K, Tomioka N, Sugita M, Yokoshi S, Nozawa A, Funakoshi K. Equation of state of $(Mg_{0.8}Fe_{0.2})_2SiO_4$ ringwoodite from synchrotron X-ray diffraction up to 20 GPa and 1700 K. Eur J Mineral 2006, 18, 523–528.
[55] Greenberg E, Dubrovinsky LS, McCammon C, Rouquette J, Kantor I, Prakapenka V, Rozenberg GK, Pasternak MP. Pressure-induced structural phase transition of the iron end-member of ringwoodite (gamma-Fe_2SiO_4) investigated by X-ray diffraction and Mossbauer spectroscopy. Am Mineral 2011, 96, 833–840.
[56] Yusa H, Inoue T, Ohishi Y. Isothermal compressibility of hydrous ringwoodite and its relation to the mantle discontinuities. Geophys Res Lett 2000, 27, 413–416.
[57] Wang J, Sinogeikin SV, Inoue T, Bass JD. Elastic properties of hydrous ringwoodite. Am Mineral 2003, 88, 1608–1611.
[58] Wang J. Elastic properties of hydrous ringwoodite at high-pressure conditions. Geophys Res Lett 2006, 33, L14308.
[59] Jacobsen SD, Smyth JR, Spetzler H, Holl CM, Frost DJ. Sound velocities and elastic constants of iron-bearing hydrous ringwoodite. Phys Earth Planet Inter 2004, 143, 47–56.
[60] Jacobsen SD, Smyth JR. Effect of water on the sound velocities of ringwoodite in the transition zone. In S.D. Jacobsen and S. Van der Lee, Eds., Earth's Deep Water Cycle 2006, 168, p. 131–145. Am Geophys Union, Geophys Monogr, Washington, D.C.
[61] Manghnani MH, Amulele G, Smyth JR, Holl CM, Shen G, Prakapenka V. Equation of state of hydrous Fo90 ringwoodite to 45 GPa by synchrotron powder diffraction. Mineral Mag 2005, 69, 317–324.
[62] Smyth JR, Holl CM, Frost DJ, Jacobsen SD. High pressure crystal chemistry of hydrous ringwoodite and water in the Earth's interior. Phys Earth Planet Inter 2004, 143, 271–278.
[63] Suzuki I, Ohtani Em Kumazawa M. Thermal expansion of γ-Mg_2SiO_4. J Phys Earth 1979, 27, 53–61.

[64] Akaogi M, Ito E, Navrotsky A. Olivine-modified spinel-spinel transitions in the system Mg_2SiO_4-Fe_2SiO_4 – calorimetric measurements, thermochemical calculation, and geophysical application. J Geophys Res 1989, 94, 15671–15685.
[65] Chopelas A. Thermal expansivity of mantle relevant magnesium silicates derived from vibrational. spectroscopy at high pressure. Am Mineral 2000, 85, 270–278.
[66] Katsura T, Yokoshi S, Song M, Kawabe K, Tsujimura T, Kubo A, Ito E, Tange Y, Tomioka N, Saito K, Nozawa A, Funakoshi K. Thermal expansion of Mg_2SiO_4 ringwoodite at high pressures. J Geophys Res 2004, 109, B12209.
[67] Yu YG, Wentzcovitch RM. Density functional study of vibrational and thermodynamic properties of ringwoodite. J Geophys Res 2006, 111, B12202.
[68] Li L, Weidner DJ, Brodholt J, Alfe D, Price GD. Elasticity of Mg_2SiO_4 ringwoodite at mantle conditions. Phys Earth Planet Inter 2006, 157, 181–187.
[69] Ottonello G, Civalleri B, Ganguly J, Zuccolini Vetuschi M, Noel Y. Thermophysical properties of the alpha-beta-gamma polymorphs of Mg_2SiO_4: a computational study. Phys Chem Miner 2009, 36, 87–106.
[70] Nishihara Y, Takahashi E, Matsukage KN, Iguchi T, Nakayama K, Funakoshi K. Thermal equation of state of $(Mg_{0.91}Fe_{0.09})_2SiO_4$ ringwoodite. Phys Earth Planet Inter 2004, 143, 33–46,
[71] Yagi T, Akaogi M, Shimomura O, Suzuki T, Akimoto S. In situ observation of the olivine-spinel phase-transformation in Fe_2SiO_4 using synchrotron radiation. J Geophys Res 1987, 92, 6207–6213.
[72] Fei YW, Saxena SK. A thermochemical database for phase-equilibria in the system Fe-Mg-Si-O at high-pressure and temperature. Phys Chem Miner 1986, 13, 311–324.
[73] Plymate TG, Stout JH. Pressure-volume-temperature behavior of gamma-Fe_2SiO_4 (spinel) based on static compression measurements at 400 °C. Phys Chem Miner 1994, 21, 413–420.
[74] Yu YG, Vinograd VL, Winkler B, Wentzcovitch RM. Phase equilibria of $(Mg,Fe)_2SiO_4$ at the Earth's upper mantle conditions from first-principles studies. Phys Earth Planet Inter 2013, 217, 36–47.
[75] Jacobs MHG, de Jong BHWS. An investigation into thermodynamic consistency of data for the olivine, wadsleyite and ringwoodite form of $(Mg,Fe)_2SiO_4$. Geochimica et Cosmochimica Acta 2005, 69, 4361–4375.

Enrico Mugnaioli

6 Investigation of bio-related minerals by electron-diffraction tomography: Vaterite, dental hydroxyapatite, and crystalline nanorods in sponge primmorphs

6.1 Introduction

The definition of 'biomineral' includes either biologically-controlled minerals (BCM), i.e. minerals synthesized by an organism for a specific function (e.g. support, motion, protection, predation, chemical storage, optical, gravitation or magnetic sensing), or biologically-induced minerals (BIM), i.e. minerals produced as a byproduct of an organism's metabolism [1, 2]. Understanding the processes that regulate biomineral formation and make them to arrange in complex tissue architectures is of primary importance for life sciences and for developing efficient diagnoses and medical treatments for diseases like osteoporosis and Alzheimer [3, 4]. In the last twenty years, biominerals raised a remarkable interest also in materials sciences, due to the growing attention for bio-inspired, bio-mimetic, and bio-functionalized materials, and for the influence of organic agents in the synthesis of anisotropic nano-materials [5, 6]. A considerable attention is dedicated to the development of materials for prosthesis and body implantations, which must be compatible with human tissues and endowed by enhanced mechanical properties and durability [7]. Moreover, biologic processes played a major role in the evolution of Earth surface and atmosphere and probably were responsible for the appearance of most of nowadays known mineralogical species [8]. Thus, paleo-biomineralization is one of the keystones for understanding not only biologic evolution, but also climatic and geologic changes.

Biominerals are generally found only as micrometric or sub-micrometric crystals. BCM in particular are required to perform very specific functions at cellular level or to be organized in hierarchical architectures made of a complex arrangement of inorganic microcrystals and organic elements [9–12]. A comprehensive characterization of biominerals is therefore hardly feasible, and often unattainable, by conventional optical microscopy and X-ray methods, even using synchrotron sources. Crystals are normally not large enough for single-crystal X-ray diffraction. In some cases, the extraction of sufficient material for X-ray powder diffraction (XRPD) from biological tissues is a serious issue and the nanometric size of the coherent crystalline domains, associated with the eventual presence of multiple mineral species (phases), may results in diffraction data affected by significant peak broadening and overlap. Moreover, the investigated materials are anyway analyzed out of their biologic environment and any information about biomineral occurrence and arrangement into the biologic tissue is lost.

On the contrary, electron-microscopy techniques can contribute considerable information about biomineral crystal-habit, structure, and arrangement in the native biologic environment. Both scanning electron microscopy (SEM) and transmission electron-microscopy (TEM) have been widely used for the characterization of vegetal, animal and human mineralized tissues [2]. TEM in particular proved to be a versatile and powerful technique able to display nanoscopic features and to work with very limited volumetric amount of sample, thus allowing the investigation of very specific biologic environments [13–15] and facilitating the study of time-related processes, for instance connected with progressive growth stages or production of short-term intermediates [16, 17].

Biologic samples for TEM study are normally obtained after a complex protocol that involves fixation, dehydration, embedding and eventually staining. The most common procedure consists in cutting thin sample slices using an ultramicrotome. More recently, focused ion beam (FIB) allowed preparing comparable slices from more specific sample areas [15]. Cryo-facilities, able to keep the sample at liquid N_2 temperature, are frequently used both for sample preparation and TEM observation, in order to reduce the heating produced by cutting and exposure to the electron beam.

The most common imaging modes are bright-field TEM, where micrographs combining density and phase contrast are normally recorded by photographic plates or charge-coupled device (CCD) cameras, and dark-field scanning-transmission electron microscopy (STEM), where the electron beam is rapidly scanning over the sample and the inelastic scattered electrons are recorded by a high-angle annular dark field (HAADF) detector. STEM contrast is given only by difference in density and is quite convenient for imaging inorganic particles against the low-density, low-contrast organic background. Additionally, STEM significantly reduces the beam dose-rate on the sample because the beam is not kept stationary.

Both TEM and STEM imaging-modes may deliver information about single particle size, shape and arrangement in the biologic tissue. TEM and STEM images can also be acquired sequentially in fixed angular steps and used for tomographic reconstructions in order to get accurate three-dimensional information about particle arrangement [18] and particle morphology, eventually up to the identification of exposed crystal facets [19]. Imaging information can be complemented by chemical composition, obtained by energy dispersive X-ray spectroscopy (EDX or EDS) and electron energy loss spectroscopy (EELS), magnetic properties, obtained by electron holography (EH) [14, 16, 18, 20] and energy-loss magnetic chiral dichroism (EMCD) [21], and crystal structure information.

High resolution transmission electron microscopy (HRTEM) has been widely used to get crystal structure information about biomineral particles. This technique has been used for identifying particle morphology [9, 22], orientation [12, 16] and mineralogical phase [4, 14]. Though, complete ab-initio structure determination of a fully unknown phase by imaging is limited by the difficulty in obtaining comprehensive three-dimensional information with atomic resolution and by the strong illumination required, which quickly degrades and destroys beam-sensitive biominerals and their organic environment.

On the other hand, electron diffraction allows the acquisition of three-dimensional single-crystal structural data from crystallites of few nanometers, i.e. 10–1000 times smaller than the crystallites normally feasible for X-ray single-crystal analysis. Remarkably, this information is acquired using a much milder illumination than the one required for HRTEM and is significantly less affected by objective lens aberrations. Despite its remarkable potential, selected area electron-diffraction (SAED) has been only used for recognizing crystal orientation [15] and distinguishing among known mineral phases [13, 23].

Convergent beam electron-diffraction (CBED) is another technique able to deliver unique information about the structure of nanomaterials. CBED is produced by a convergent electron beam, focused on a very tiny area of the sample. Dynamical scattering is responsible for the appearance of interference bands (Kikuchi lines), while diffraction spots are replaced by diffraction disks, each of them displaying internal extinction patterns. The Friedel law is violated by dynamical scattering and hence the real symmetry of the crystal can be determined by combining the symmetry of transmitted beam, first order Laue zone and higher order zones [24, 25]. Despite this remarkable ability, CBED found as well a limited use in the study of biomineralizations [26–28].

The little use of electron diffraction for the study of biominerals, and of nanomaterials in general, is connected with the traditional limits of conventional electron-diffraction data recorded by orienting well-recognizable, low index crystallographic zones. This data acquisition strategy limits the number of sampled reflections and emphasizes dynamical scattering that ruffle reflection intensities [29, 30]. Additionally, obtaining a satisfactorily ample collection of well-aligned crystallographic zones from a nanoparticle requires an expert microscope user and a certain timeframe that may be critical for beam sensitive materials like organic samples [31].

The need for less dynamical and more complete data sets led to the development of the automated diffraction tomography (ADT) method [32–35]. ADT data collection strategy and analysis allows for more complete, less dynamical and faster electron-diffraction data collection. In this chapter, we review three case studies of bio-related minerals structurally characterized by ADT method [36–38], emphasizing ADT potential for the investigation of complex nanocrystalline materials, eventually sampled inside their biologic native environment.

6.2 Automated diffraction tomography (ADT)

ADT data collection consists in the acquisition of a series of electron-diffraction patterns while the sample is rotated around an arbitrary tilt axis by sequential mechanical steps and within the full tilt range of the microscope goniometer (Fig. 6.1a) [32, 35]. The tilt axis generally does not correspond to a specific crystallographic direction. Data

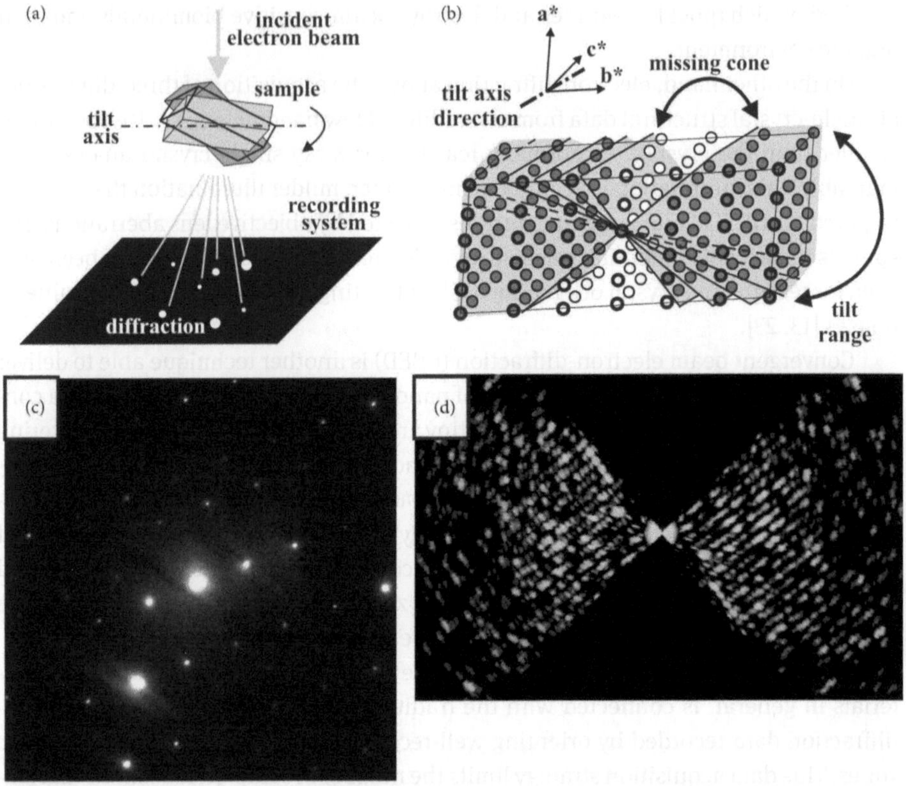

Fig. 6.1: ADT data collection and analysis. (a) Scheme showing ADT acquisition. The crystal is tilted around an arbitrary tilt axis while sequential off-zone diffraction patterns are recorded. (b) Reciprocal space sampling during ADT acquisition. The tilt axis does not correspond to any prominent crystallographic axis of the structure. All reflections following in the tilt range (grey) are sampled, even if not belonging to low index crystallographic axes. Tilt range depends on the TEM construction and on the sample holder. Reflections following in the missing cone wedge (white) cannot be sampled. (c) An exemplary single off-zone diffraction-pattern collected by ADT method. Reflections cannot be easily recognized and indexed on the basis of a single pattern. (d) A three-dimensional diffraction volume reconstructed on the basis of ADT data viewed along the tilt axis of the acquisition.

completeness is higher than for conventional in-zone data, as the whole reciprocal space available within the mechanical limits of the microscope goniometer is sampled, including those reflections not belonging to low-index zones (Fig. 6.1b).

ADT diffraction data are collected in nano-electron diffraction (NED) mode, using mild illumination conditions obtained by high condenser spot size settings and a 10 μm objective aperture. The probed area can be as small as 40–50 nm in diameter (a smaller probe introduces significant geometrical distortions in the diffraction patterns that hamper the three-dimensional reconstruction). Crystal position is tracked after each mechanical tilt step by scanning transmission electron microscopy (STEM), keeping similar lens current settings. The electron dose-rate is estimated at 3–130 e/Å^2 per second [32] and the total exposition time may be less than 2 minutes for a complete data acquisition (0.5–2 seconds exposure per about 120 patterns). Despite the low electron dose-rate, some materials, like organics and biological tissues, are still affected by beam damage and may not stand for a full acquisition tilt. In such cases, cooling the sample by a cryo-holder significantly increases its stability under the beam.

Differently from traditional in-zone electron-diffraction patterns, ADT non-oriented patterns cannot be directly indexed and measured (Fig. 6.1c). Though, because each pattern is acquired off-zone and fewer reflections are simultaneously excited, multiple scattering and dynamical effects are significantly reduced. After the acquisition, the recorded stack of two-dimensional non-oriented patterns must be transformed into a three-dimensional diffraction volume (Fig. 6.1d). On the basis of this reconstruction, cell vectors, containing cell parameter and orientation information, are automatically defined by clustering routines working in difference vector space [33, 39] or arbitrarily defined on the basis of visual inspection. Cell parameters, even for triclinic lattices, may be determined with an accuracy of 1–2 %. Local structural features like disorder, twinning and polycrystallinity can be also visualized in the three-dimensional space and related to cell parameters and crystal morphology.

Reflections sampled by the ADT method are generally affected by a non-predictable excitation error. i.e. reflection intensities are under-sampled if the Ewald sphere does not cut the middle of the reflection [40]. In order to improve reflection intensity data, each ADT pattern can be acquired in precession electron-diffraction (PED) mode [34, 41]. The electron beam is slightly tilted with respect to the vertical optical axis and rapidly rotated around it. In such a way, instead of recording a stationary cut of the reciprocal space, it is possible to perform an integration over a portion of the reciprocal space and over each reflection volume. Actually, PED guarantees a significant reduction of the influence of excitation error on experimental intensities recorded by the ADT method [34].

The internal residual of ADT intensity data-sets is normally in the range of 15–25 % due to the experimental instability of goniometer rotation, residual excitation error and dynamical effects, missing routines for correcting absorption and geometrical effects. This inaccuracy results in relatively high structure residuals that hamper a proper kinematical structure refinement. Nonetheless, ADT data sets were successfully used for ab-initio structure determination of several unknown nanocrystalline materials [42], like minerals [43–45], zeolites [46–48] intermetallic nanoparticles [49] and organic compounds [31, 50].

6.3 Experimental

6.3.1 Conventional TEM, ADT and structure analysis

TEM, STEM, EDX, and ADT were performed with a *FEI* TECNAI F30 STWIN transmission electron-microscope equipped with field emission gun and working at 300 kV. TEM imaging and diffraction were recorded by a CCD camera (14-bit 1024×1024 pixel *GATAN* 794MSC). STEM imaging was recorded by a *FISHIONE* HAADF detector. Samples were mounted on a tomographic *FISCHIONE* holder capable of a tilt range up to $\pm 60°$.

ADT was executed by a semi-automatic software able to perform sequentially crystal tilt, crystal position tracking and diffraction pattern acquisition [32]. Each data collection was performed in tilt steps of 1°. PED was performed with a *NANOMEGAS* DigitStar device keeping the precession angle at 1.2°.

ADT data elaboration, including three-dimensional diffraction volume reconstruction and visualization, cell parameter determination and reflection intensity integration, was carried out via *ADT3D* software and in-home developed routines [33–35, 39]. Ab-initio structure determination for vaterite and dental hydroxyapatite were done using direct methods implemented in the software *SIR2008* [51] and *SIR2011* [52]. Structure refinement was done by *SHELX97* [53]. The kinematical approximation $I_{hkl} \sim F_{hkl}^2$ was always assumed and no correction for geometrical effects or absorption was applied.

6.3.2 Sponge primmorphs sample preparation

Live specimens of *Suberites domuncula* (Porifera, Demospongiae, Hadromerida) were collected in the Adriatic Sea near Rovinj (Croatia) and kept in artificial sea-water aquaria for over ten months. Single primmorphs cells were obtained with the procedure described in Müller et al. (1999) [54] and Mugnaioli et al. (2009) [36]. After staining, dehydration and fixation, the sample was finally fixed in Araldite and sliced using ultramicrotome *Leica* Ultracut S.

6.3.3 Vaterite synthesis

Two samples of vaterite were obtained by different synthetic routes. Standard spherical vaterite aggregates were prepared by aqueous synthesis as described in Dandeu et al. (2006) [55] and Mugnaioli et al. (2012) [37]. The sample was thereafter crushed with a *FRITSCH* Pulverisette 7 ball mill equipped with ZrO_2 balls. Single-phase vaterite nanoparticles were obtained by the non-aqueous synthesis described by Schüler & Tremel (2011) [56].

6.3.4 Human hydroxyapatite extraction

Human dental enamel and dentine powder were taken from permanent human molar and premolar teeth from persons of 25 to 35 years of age of both sexes. Enamel and dentine were mechanically separated using a hand-guided dental drill. After separation, both samples were milled in an agate mortar [38].

6.4 Characterization of unknown biominerals inside tissues: Crystalline nanorods in sponge primmorphs

Sponges are among the most ancient multicellular organisms. They have an endoskeleton composed by thousands of structural elements named spicules that may be made of spongine, calcium carbonate or amorphous silica (SiO_2). During the investigation of the first stages of siliceous spicule formation, sponge primmorph-cells from *Suberites domuncula* (Porifera, Demospongiae, Hadromerida) were cultivated in vitro and subsequently sliced for electron-microscopy observations. STEM imaging revealed the presence of rod-like structures much brighter than the surrounding organic environment [36]. The rods had a typical size of 300–600 nm in length and 20–50 nm in width and were commonly observed intracellularly, hosted in specific vesicles (Fig. 6.2a-d).

The average atomic composition of the nanorods, detected by EDX, was (atomic %): O 63.5, Si 21.2, Al 11.5, Mg 1.9, K 0.9, Fe 0.9. Crystallinity and major presence of Al and other elements clearly distinguished these nanorods from mature spicules made of amorphous silica. HRTEM showed that nanorods were completely or partially crystalline, with a layered structure characterized by ~ 10 Å interlayer spacing (Fig. 6.2e) parallel to the growth direction of the rod [36]. Nonetheless, it was not possible to get a complete structure characterization of the nanorods by imaging, because nanorods were always characterized by a pervasive defectivity, like fragmentation, kinking, and bending (Fig. 6.2c–d). Moreover, the sample could stand HRTEM illumination-conditions only for few minutes.

Three-dimensional ADT data taken from single nanorods allowed to unambiguously determine a *C*-centered cell with parameters $a = 5.2$ Å, $b = 9.2$ Å, $c \sim 22$ Å, $\beta \sim 95°$ (Fig. 6.2f). Cell parameters and composition are consistent with layered silicates (phyllosilicates). The not perfect crystallinity of the sample determined a high variability of the *c* parameter and did not allow for a more precise identification of the mineral phase that is probably an interstratified clay mineral.

Interestingly, nanorods are commonly associated with other high-contrast features, like rounded high-contrast areas enriched in amorphous silica (Fig. 6.2b), Fe-rich nodules of few tens of nanometers interpreted as ferritin cores [57, 58] (Fig. 6.2b) and amorphous nanofibril bundles enriched in silica and possibly associated with the first stages of axial filament formation [59] (Fig. 6.2d). The role of all these intra-

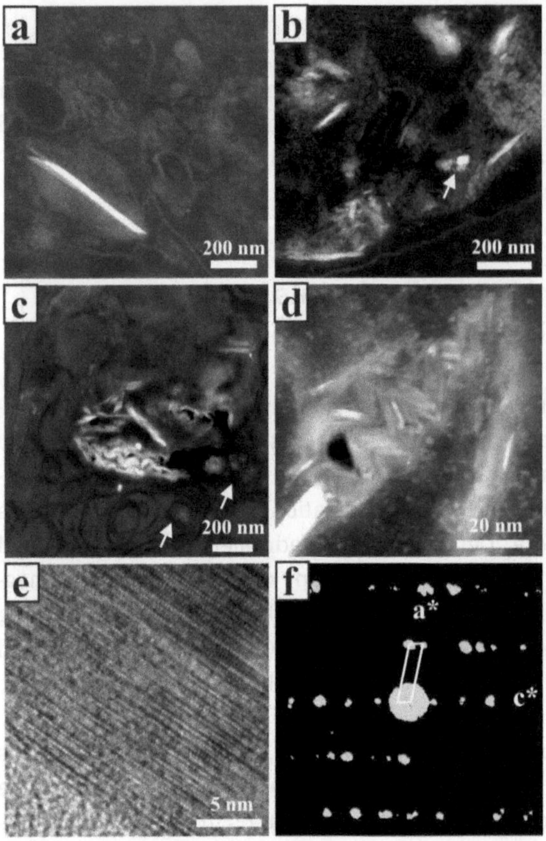

Fig. 6.2: Crystalline inorganic nanorods in sponge primmorphs. (a) Dark-field STEM image of a nanorod inside a vesicle located intracellularly. (b) Dark-field STEM image of a vesicle hosting nanorods, large rounded areas enriched in silica and a small nodule enriched in iron (marked by arrow). (c) Dark-field STEM image of a vesicle hosting kinked and bent nanorods. In the neighborhood some fibril bundles enriched in silica are marked by arrows. (d) Dark-field STEM image of a vesicle hosting a kinked nanorod with incoherent crystalline layers. (e) Bright-field HRTEM image of the 'almost' coherent crystalline nanorod shown in panel (a). (f) ADT three-dimensional reconstructed diffraction volume for the nanorod showed in panel (a), viewed along b* direction. The unit cell is sketched in white.

cellular features during sponge spiculogenesis is still not fully understood. Crystalline nanorods may play a role during the early stages of axial filament formation acting as a rigid template, or they may be floating clay crystallites taken up from the sea water and used as a silica source for spicules after dissolution and removal of undesired elements [59]. Nevertheless, ADT was the only method able to deliver unambiguous unit-cell information about nanorods, i.e. about fully unknown and isolated nanoscopic crystals directly sampled inside a biologic tissue.

6.5 Ab-initio structure determination of complex bio-related nanocrystalline minerals: Vaterite

Calcium carbonate is by far the most abundant among bio-related compounds. Among other occurrences, $CaCO_3$ polymorphs (calcite, aragonite and vaterite) are found as BCM in the skeleton of coccolites, foraminifera and mollusks, in avian eggshells, crab cuticole, sponge spicules, fish gravity receptors, coral and echinoderm shells, and as calcium reservoir in plants [2]. In addition, quantities of calcium carbonate precipitates are continuously produced in aquatic environment by algae and cyanobacteria activity [2].

While calcite and aragonite commonly form up to centimetric crystals and their structure is well known from X-ray crystallography, vaterite exists only in form of nanoscopic crystals, typically forming polycrystalline spherical aggregates (Fig. 6.3a). As a consequence, the atomic structure of vaterite is still debated, despite its widespread occurrence as mineral in Nature and its primary role in biomineralization, weathering, and nucleation processes [60–62]. The five experimental models hitherto proposed for vaterite on the basis of X-ray diffraction [63–67] comprise hexagonal or orthorhombic symmetry. The corresponding unit cells are commensurate with the smallest cell proposed by Kamhi (1963) [64], and therefore difficult to distinguish on the basis of the low-quality diffraction data available. Though, none of these models is able to explain all the physical and spectroscopic observations [37].

Fig. 6.3: Synthetic vaterite nanocrystals. (a) Dark-field STEM image of typical spherical aggregates of vaterite. A zoom of the area marked by the grey rectangle is shown in the inset, in order to point out the needle-like morphology of vaterite nanocrystals. (b) Bright-field HRTEM image of a vaterite nanoparticle characterized by disorder and local superstructure sequences.

The ADT method allowed for the first time ab-initio structure determination of vaterite by three-dimensional single-crystal data taken from single crystallites of about 50 nm [37]. ADT revealed a *C*-centered monoclinic cell of a = 12.17 Å, b = 7.12 Å, c = 9.47 Å, β = 118.95°. Despite the monoclinic symmetry, this cell corresponds to

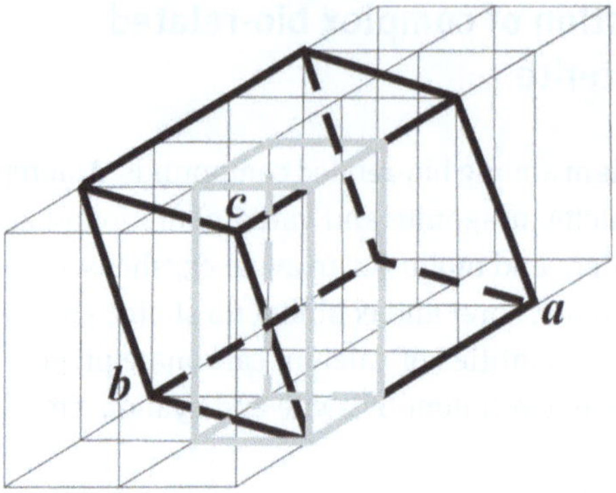

Fig. 6.4: Vaterite cell (black thick lines) determined on the basis of ADT data [37], superimposed to Khami's cell [64] (grey lines).

a superstructure of Kamhi's cell [64] (Fig. 6.4). The vaterite structure was determined ab-initio in space group *C2/c* by direct methods and refined imposing flat CO_3 groups and restraints on C-O and O-O interatomic distances (Fig. 6.5). Remarkably, this structure is isotypic with the high-temperature polymorph of $(Y_{0.92}Er_{0.08})BO_3$, previously described as a vaterite-like structure on the basis of the powder-diffraction profile and recently independently solved in monoclinic symmetry [68].

The so-determined vaterite structure shows an unusual stacking of cationic and anionic layers along the [001] direction. Ca^{2+} ions are packed in pseudo-hexagonal layers parallel to (001), which alternate with layers of CO_3^{2-} groups arranged in a bridging mode. The main difference with previously proposed vaterite structures is a reciprocal shift of anionic layers that destroys the hexagonal symmetry and splits both Ca and C sites in two crystallographically distinct positions. This model can explain

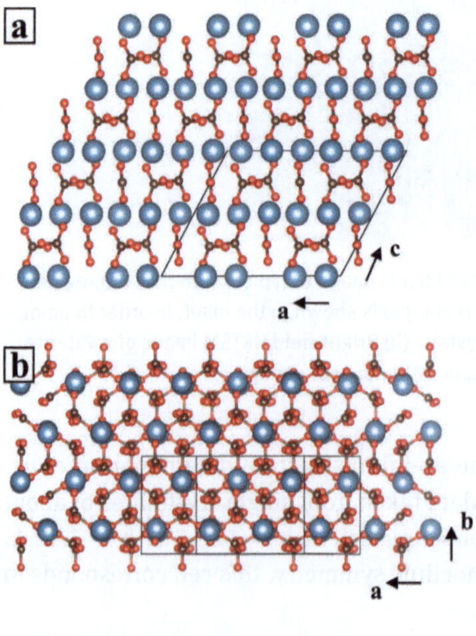

Fig. 6.5: Vaterite atomic structure determined and refined on the basis of ADT data. ***a***) View along **b**. ***b***) View along the pseudo-hexagonal axis, approximately corresponding to the crystallographic direction [103]. Big blue spheres: calcium atoms; small brown spheres: carbon atoms; small red spheres: oxygen atoms.

spectroscopic and physical observations significantly better than previous structure models (see Mugnaioli et al. (2012) [37] for the complete discussion).

Recently, Wang and Becker (2009) [69] and Demichelis et al. (2012) [70] proposed different vaterite models on the basis of ab-initio calculations. These models can also account for most of spectroscopic evidences. On the other hand, Wang et al. (2012) [71] showed that their calculations may lead also to a monoclinic *C*2/*c* model, thus independently achieving a structure very similar to the one determined experimentally on the basis of ADT data [37]. Also Demichelis et al. (2013) [72] showed that the monoclinic model, eventually relaxed to the acentric space group *C*2, is almost isoenergetic with their trigonal model calculated ab-initio [70]. In fact, the three models proposed by Mugnaioli et al. [37], Wang and Becker [69], and Demichelis et al. [70] differ only for the stacking sequence of anionic layers and cannot be clearly distinguished on the basis of experimental evidences like XRPD, HRTEM, SAED [73] and Raman spectroscopy [74]. Actually, the problem of establishing a conclusive vaterite structure is further complicated by the pervasive stacking disorder already observed at the nanoscale and by the occurrence of local, more complex periodic or quasi-periodic modulations [37] (Fig. 6.3b). Moreover, other authors provided evidence suggesting that synthetic and biologic vaterite have different structures, and that even different varieties of both synthetic and biologic vaterite exist [75, 76].

ADT allowed determining ab-initio the structure of vaterite, or at least of its basic structure motif, from three-dimensional experimental electron-diffraction data obtained from low-defective single crystals with a size of about 50 nm. This model was not able to close the discussion about the vaterite structure, but pushed the research toward lower symmetry that had been previously neglected. This model also consolidated the evidence that the vaterite structure is characterized by a marked separation between anionic and cationic layers leading to very unfavorable Coulomb contributions to the lattice energy. This may explain why vaterite is the first $CaCO_3$ polymorph to form in solution and why it is confined to a nanometric size, progressively replaced by calcite or aragonite during crystal growth [77].

6.6 Structure investigation of human calcified tissues: Dental hydroxyapatite

Animal calcified tissues, like mollusk nacre [12] or vertebrate bones and teeth [78], consist of nanocrystalline inorganic particles and organic elements organized in a hierarchical architecture over several length scales. This complex arrangement allows tuning physical and chemical properties in order to absolve specific biological functions [10, 11]. Macroscopic properties are thus regulated by size, arrangement, atomic structure, and growth evolution of microscopic and sub-microscopic elements.

Human calcified tissues absolve the main functions of structural support (bones) and food mechanical grinding (tooth enamel and dentine), and at the same time they are the main reservoirs of calcium and phosphorous for the human body [78]. The atomic structure of bones and teeth nanocrystals, determined by X-ray diffractometry, and infra-red and Raman spectrometry, is very similar to that of geologic and synthetic hydroxyapatite, $Ca_6Ca_4(PO_4)_6(OH)_2$, which crystallizes in space group $P6_3/m$ [79]. The mirror plane orthogonal to the unique **c** axis is preserved by a statistical positioning (½ occupancy) of OH^- groups along the [00*l*] channels (Fig. 6.6a). When different anionic species enter the hydroxyapatite structure, OH^- groups may get ordered and correspondingly the symmetry reduces to the monoclinic space group $P2_1/b$ (Fig. 6.6b) [80]. Elliot et al. (1973) [81] proposed that hexagonal and monoclinic hydroxyapatite polymorphs coexist in human calcified tissues.

Dental enamel and dentine nanocrystals were recently investigated by TEM imaging, ADT, and CBED [38]. Teeth tissues are composed by elongated nanocrystals with platelet- or needle-like habit and size typically smaller than 100 nm, typically tile-piled in 'bundles' a few microns long [Fig. 6.7a-b]. Five independent ADT data sets, three for dental enamel and two for dentine, were acquired from crystals of about 50 nm in size. All ADT volumes (Fig. 6.7c) were equivalent and showed single-crystal hexagonal lattices with no extra reflections typical of the monoclinic modification [80, 81]. The crystallographic direction **c** was always parallel to the main direction of crystal growth. Unexpectedly, ab-initio structure determination and refinement in acentric space group $P6_3$ resulted in significant lower structural residuals and clearer potential maps when compared with the related centrosymmetric space group $P6_3/m$. The main difference between centrosymmetric and acentric structures concerns the OH^- group, as in $P6_3$ a single well-defined OH^- position breaks the mirror plane (Fig. 6.6c).

The acentric symmetry of the dental crystals was confirmed by CBED analysis of the same enamel sample (Fig. 6.7d) [38]. Patterns taken along the zone [11–20] clearly showed a mirror plane along **c***, while no mirror plane was observed orthogonal to **c***. This symmetry is consistent with the acentric space group $P6_3$. Previous CBED analyses of human hydroxyapatite mostly focused on higher order Laue-zones [26]. The difference between acentric and centrosymmetric model is anyway confined to the OH^- group position, and this may produce little evidence on these patterns.

According to these evidences, Mugnaioli et al. (2014) [38] considered that the atomic structure of human dental nanocrystals is more reasonably described in the acentric space group $P6_3$. Enamel and dentine tissues would therefore consist of hierarchically arranged acentric nanocrystals, characterized by an electrostatic potential along the unique **c** axis, i.e. the main growth direction. Electrostatic potential in acentric crystals may be tuned by stress and temperature, while the application of an external electric force determines anisotropic stress and strain. This experimental evidence is of primary importance for understanding the influence of electric fields and the role of osteoblast cells in the morphogenesis process of calcified tissues [82, 83]. Remarkably, it was recently reported that also thin films of synthetic hydroxyapatite show acentric physical properties [84].

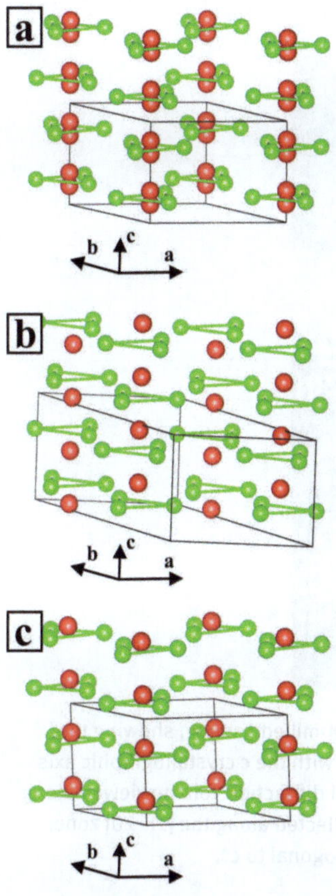

Fig. 6.6: Hydroxyapatite structures. Only the oxygen atoms belonging to OH^- groups and the neighboring calcium atoms are shown. (a) Pure geological hydroxyapatite, space group $P6_3/m$. (b) Hydroxyapatite with partially substituted anionic species, space group $P2_1/b$. **c)** Human dental nanocrystal structure as determined and refined on the basis of ADT data, space group $P6_3$. Red spheres: oxygen; green spheres: calcium. Note that in (a) OH^- position is split along [00*l*] channels, in (b) is alternated in different channels, in (c) is always shifted in the same direction.

6.7 Conclusions and perspectives

Macroscopic properties of biologic mineralized tissues are regulated by a complex hierarchical architecture made of nanocrystalline inorganic BCM and organic elements. TEM offers a wide range of techniques for characterizing mineralized tissues through imaging, holography, chemical spectroscopy, and diffraction [2]. In this regard, the ADT method has the remarkable ability to collect three-dimensional quasi-kinematical atomic structure information from single nanograins directly inside their native biologic environment. At the same time, ADT is very suitable also for the characterization of nanocrystalline BIM aggregates produced as byproduct of biologic activities. Compared with X-ray techniques, electron diffraction gives the possibility to spot minority phases and isolated nanodomains in polyphasic or polymorphic mixtures and to get unambiguous structural information from phases characterized by long cell parameters, low crystallinity, and pseudo-symmetry [35, 42, 47].

The success of ADT has encouraged other groups to develop their own systems for tomographic acquisition of electron-diffraction data [85, 86]. Electron-diffraction to-

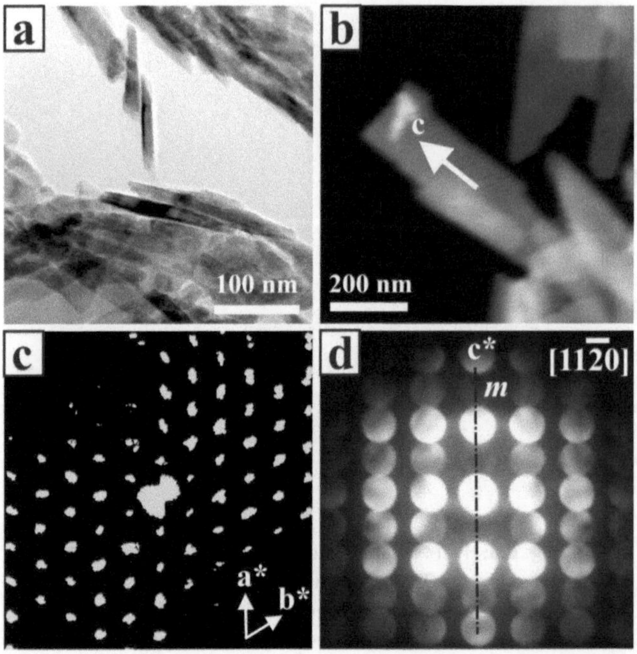

Fig. 6.7: Dental hydroxyapatite. (a) Bright-field TEM image of a non-milled sample, showing tile-piled crystals. (b) Dark-field STEM image of a platelet-like crystal, with the **c** crystallographic axis parallel to the main direction of growth. (c) ADT three-dimensional diffraction volume viewed along **c*** and showing the hexagonal cell symmetry. (d) CBED pattern collected along the [11–20] zone, showing a mirror plane (*m*) parallel to **c***, but no mirror plane orthogonal to **c***.

mographic acquisition has been coupled with electron- beam tilt for a finer reciprocal space sampling (rotation electron-diffraction, RED [87, 88]) and with energy filter for removing the amorphous and inelastic scattering [89]. Additionally, software able to use ADT data for dynamical structure refinement has been recently proposed [90, 91].

In the next future, the empowering of aberration-corrected TEM [92] will allow to further reduce the sampled area and electron dose-rate during imaging and electron-diffraction data collection, while environmental TEM will offer the possibility to analyze functioning biologic cells [93]. Nannenga et al. (2014) [94] already demonstrated that tomographic electron-diffraction data may be used for the structure determination of proteins available only in crystals of few microns. Allegedly, the availability of very sensitive direct electron detectors [95] will soon allow using electron-diffraction tomography not only for characterizing inorganic biominerals, but also for getting structural insights about organic components of biologic tissues.

Acknowledgment: The author is thankful to Ute Kolb, Iryna Andrusenko and all other researchers that collaborated in the characterization of biominerals by ADT. This work was supported by the Italian project FIR2013 – Exploring the Nanoworld.

Bibliography

[1] Mann S. Biomineralization: Principles and concepts in bioinorganic materials chemistry. Oxford, UK, Oxford University Press, 2001.
[2] Pósfai M, Kasama T, Dunin-Borkowski RE. Biominerals at the nanoscale: transmission electron microscopy methods for studying the special properties of biominerals. In: Nieto F, Livi KJT, eds. Minerals at the Nanoscale. London, UK, Mineralogical Society of Great Britain and Ireland, 2013, 377–435.
[3] Glimcher MJ. Bone: Nature of the calcium phosphate crystals and cellular, structural, and physical chemical mechanisms in their formation. In: Sahai N, Schoonen, MAA, eds. Medical mineralogy and geochemistry. Chantilly, VA, USA, Mineralogical Society of America, 2006, 223–382.
[4] Quintana C, Cowley JM, Marhic C. Electron nanodiffraction and high-resolution electron microscopy studies of the structure and composition of physiological and pathological ferritin. J Struct Biol 2004, 147, 166–178
[5] Dujardin E, Mann S. Bio-inspired materials chemistry. Adv Mater 2002, 14, 775–788.
[6] Nudelman F, Sommerdijk NAJM. Biomineralization as an inspiration for materials chemistry. Angew Chem Int Edit 2012, 51, 6582–6596.
[7] Sammons RL, Marquis PM, Macaskie LE, Yong P, Basner C. Developments in the use of calcium phosphates as biomaterials. In: Valsani-Jones E, ed. Phosphorus in environmental technology: principles and applications. London, UK, IWA Publishing, 2004, 582–609.
[8] Hazen RM, Papineau D, Bleeker W et al. Mineral evolution. Am Mineral 2008, 93, 1693–1720.
[9] Mann S, Sparks NHC. Single crystalline nature of coccolith elements of the marine alga *Emiliania huxleyi* as determined by electron diffraction and high-resolution transmission electron microscopy. P Roy Soc Lond B 1988, 234, 441–453.
[10] Currey JD. Hierarchies in biomineral structures. Science 2005, 309, 253–254.
[11] Dunlop JWC, Weinkamer R, Fratzl P. Artful interfaces within biological materials. Mater Today 2011, 14, 70–78.
[12] Xie L, Wang X-X, Li J. The SEM and TEM study on the laminated structure of individual aragonitic nacre tablet in freshwater bivalve *H. cumingii Lea* shell. J Struct Biol 2010, 169, 89–94.
[13] Schultheiss-Grassi PP, Wessiken R, Dobson J. TEM investigations of biogenic magnetite extracted from the human hippocampus. Biochim Biophys Acta 1999, 1426, 212–216.
[14] Pósfai M, Moskowitz BM, Arató B et al. Properties of intracellular magnetite crystals produced by *Desulfovibrio magneticus* strain RS-1. Earth Planet Sc Lett 2006, 249, 444–455.
[15] Suzuki M, Kameda J, Sasaki T, Saruwatari K, Nagasawa H, Kogure T. Characterization of the multilayered shell of a limpet, *Lottia kogamogai* (Mollusca: Patellogastropoda), using SEM–EBSD and FIB–TEM techniques. J Struct Biol 2010, 171, 223–230.
[16] Simpson ET, Kasama T, Pósfai M, Buseck PR, Harrison RJ, Dunin-Borkowski RE. Magnetic induction mapping of magnetite chains in magnetotactic bacteria at room temperature and close to the Verwey transition using electron holography. J Phys Conf Ser 2005, 17, 108–121.
[17] Uebe R, Junge K, Henn V et al. The cation diffusion facilitator proteins MamB and MamM of *Magnetospirillum gryphiswaldense* have distinct and complex functions, and are involved in

magnetite biomineralization and magnetosome membrane assembly. Mol Microbiol 2011, 82, 818–835.

[18] Kasama T, Pósfai M, Chong RKK et al. Magnetic properties, microstructure, composition, and morphology of greigite nanocrystals in magnetotactic bacteria from electron holography and tomography. Am Mineral 2006, 91, 1216–1229.

[19] Buseck PR, Dunin-Borkowski RE, Devouard B et al. Magnetite morphology and life on Mars. P Natl Acad Sci USA 2001, 98, 13490–13495.

[20] Dunin-Borkowski RE, McCartney MR, Frankel RB, Bazylinski DA, Pósfai M, Buseck PR. Magnetic microstructure of magnetotactic bacteria by electron holography. Science 1998, 282, 1868–1870.

[21] Stöger-Pollach M, Treiber CD, Resch GP, Keays DA, Ennen I. EMCD real space maps of *Magnetospirillum magnetotacticum*. Micron 2011, 42, 456–460.

[22] Lefèvre CT, Pósfai M, Abreu F, Lins U, Frankel RB, Bazylinski DA. Morphological features of elongated-anisotropic magnetosome crystals in magnetotactic bacteria of the *Nitrospirae* phylum and the *Deltaproteobacteria* class. Earth Planet Sc Lett 2011, 312, 194–200.

[23] McClean RG, Schofield MA, Kean WF et al. Botanical iron minerals: correlation between nanocrystal structure and modes of biological self-assembly. Eur J Mineral 2001, 13, 1235–1242.

[24] Buxton BF, Eades JA, Steeds JW, Rackham GM. The symmetry of electron diffraction zone axis patterns. Philos T Roy Soc A 1976, 281, 171–194.

[25] Champness PE. Convergent beam electron diffraction. Min Mag 1987, 51, 33–48.

[26] Brès EF, Cherns D, Vincent R, Morniroli J-P. Space-group determination of human tooth-enamel crystals. Acta Crystallogr B 1993, 49, 56–62.

[27] Sánchez-Pastenes E, Reyes-Gasga J. Determination of the point and space groups for hydroxyapatite by computer simulation of CBED electron diffraction patterns. Rev Mex Fis 2005, 51, 525–529.

[28] Saruwatari K, Akai J, Fukumori Y, Ozaki N, Nagasawa H, Kogure T. Crystal orientation analyses of biominerals using Kikuchi patterns in TEM. J Miner Petrol Sci 2008, 103, 16–22.

[29] Dorset DL. Structural electron crystallography. New York, NY, USA, Plenum Press, 1995.

[30] Cowley JM, Goodman P, Vainshtein BK, Zvyagin BB, Dorset DL. Electron diffraction and electron microscopy in structure determination. In: Shmueli U, ed. International tables for crystallography, Volume B: Reciprocal space. 2nd ed. Dordrecht, The Netherlands, Kluwer Academic, 2001, 276–326.

[31] Kolb U, Gorelik TE, Mugnaioli E, Stewart A. Structural characterization of organics using manual and automated electron diffraction. Polym Rev 2010, 50, 385–409.

[32] Kolb U, Gorelik T, Kübel C, Otten MT, Hubert D. Towards automated diffraction tomography: Part I – Data acquisition. Ultramicroscopy 2007, 107, 507–513.

[33] Kolb U, Gorelik T, Otten MT. Towards automated diffraction tomography. Part II – Cell parameter determination. Ultramicroscopy 2008, 108, 763–772.

[34] Mugnaioli E, Gorelik T, Kolb U. “Ab initio” structure solution from electron diffraction data obtained by a combination of automated diffraction tomography and precession technique. Ultramicroscopy 2009, 109, 758–765.

[35] Kolb U, Mugnaioli E, Gorelik TE. Automated electron diffraction tomography – a new tool for nano crystal structure analysis. Cryst Res Technol 2011, 46, 542–554.

[36] Mugnaioli E, Natalio F, Schloßmacher U, Wang X, Müller WEG, Kolb U. Crystalline nanorods as possible templates for the synthesis of amorphous biosilica during spicule formation in Demospongiae. ChemBioChem 2009, 10, 683–689.

[37] Mugnaioli E, Andrusenko I, Schüler T et al. Ab initio structure determination of vaterite by automated electron diffraction. Angew Chem Int Edit 2012, 51, 7041–7055.

[38] Mugnaioli E, Reyes-Gasga J, Kolb U, Hemmerlé J, Brès ÉF. Evidence of noncentrosymmetry of human tooth hydroxyapatite crystals. Chem-Eur J 2014, 20, 6849–6852.
[39] Schlitt S, Gorelik TE, Stewart AA, Schömer E, Raasch T, Kolb U. Application of clustering techniques to electron diffraction data: determination of unit-cell parameters. Acta Crystallogr A 2012, 68, 536–546.
[40] Williams DB, Carter CB. Transmission electron microscopy, Volume II: Diffraction. New York, NY, USA, Plenum Press, 1996.
[41] Vincent R, Midgley PA. Double conical beam-rocking system for measurement of integrated electron diffraction intensities. Ultramicroscopy 1994, 53, 271–282.
[42] Mugnaioli E. Single nano crystal analysis using automated electron diffraction tomography. Rend Lincei-Sci Fis, 2015 26, 211–223.
[43] I Rozhdestvenskaya I, Mugnaioli E, Czank M et al. The structure of charoite, $(K,Sr,Ba,Mn)_{15\text{-}16}(Ca,Na)_{32}[(Si_{70}(O,OH)_{180})](OH,F)_{4.0}\bullet nH_2O$, solved by conventional and automated electron diffraction. Mineral Mag 2010, 74, 159–177.
[44] Gemmi M, Fischer J, Merlini M et al. A new hydrous Al-bearing pyroxene as a water carrier in subduction zones. Earth Planet Sc Lett 2011, 310, 422–428.
[45] Capitani GC, Mugnaioli E, Rius J et al. The Bi sulfates from the Alfenza Mine, Crodo, Italy: An automatic electron diffraction tomography (ADT) study. Am Mineral 2014, 99, 500–510.
[46] Jiang J, Jorda JL, Yu J et al. Synthesis and structure determination of the hierarchical meso-microporous zeolite ITQ-43. Science 2011, 333, 1131–1134.
[47] Mugnaioli E, Kolb U. Applications of automated diffraction tomography (ADT) on nanocrystalline porous materials. Micropor Mesopor Mat 2013, 166, 93–101.
[48] Mugnaioli E, Kolb U. Structure characterization of nanocrystalline porous materials by tomographic electron diffraction. Z Kristallogr 2015, 230, 271–288.
[49] Birkel CS, Mugnaioli E, Gorelik T, Kolb U, Panthöfer M, Tremel W. Solution synthesis of a new thermoelectric $Zn_{1+x}Sb$ nanophase and its structure determination using automated electron diffraction tomography. J Am Chem Soc 2010, 132, 9881–9889.
[50] Gorelik TE, van de Streek J, Kilbinger AFM, Brunklaus G, Kolb U. *Ab-initio* crystal structure analysis and refinement approaches of oligo *p*-benzamides based on electron diffraction data. Acta Crystallogr B 2012, 68, 171–181.
[51] Burla MC, Caliandro R, Camalli M et al. *IL MILIONE*: a suite of computer programs for crystal structure solution of proteins. J Appl Crystallogr 2007, 40, 609–613.
[52] Burla MC, Caliandro R, Camalli M et al. *SIR2011*: a new package for crystal structure determination and refinement. J Appl Crystallogr 2012, 45, 357–361.
[53] Sheldrick GM. A short history of *SHELX*. Acta Crystallogr A 2008, 64, 112–122.
[54] Müller WEG, Wiens M, Batel R et al. Establishment of a primary cell culture from a sponge: primmorphs from *Suberites domuncula*. Mar Ecol-Prog Ser 1999, 178, 205–219.
[55] Dandeu A, Humbert B, Carteret C, Muhr H, Plasari E, Bossoutrot J-M. Raman spectroscopy – A powerful tool for the quantitative determination of the composition of polymorph mixtures: Application to $CaCO_3$ polymorph mixtures. Chem Eng Technol 2006, 29, 221–225.
[56] Schüler T, Tremel W. Versatile wet-chemical synthesis of non-agglomerated $CaCO_3$ vaterite nanoparticles. Chem Commun 2011, 47, 5208–5210.
[57] Natalio F, Mugnaioli E, Wiens M et al. Silicatein-mediated incorporation of titanium into spicules from the demosponge *Suberites domuncula*. Cell Tissue Res 2010, 339, 429–436.
[58] Natalio F, Wiese S, Friedrich N, Werner P, Tahir MN. Localization and characterization of Ferritin in Demospongiae: A possible role on spiculogenesis. Mar Drugs 2014, 12, 4659–4676.
[59] Müller WEG, Mugnaioli E, Schröder HC et al. Hierarchical composition of the axial filament from spicules of the siliceous sponge *Suberites domuncula*: from biosilica-synthesizing

nanofibrils to structure- and morphology-guiding triangular stems. Cell Tissue Res 2013, 351, 49–58.
[60] Gebauer D, Völkel A, Cölfen H. Stable prenucleation calcium carbonate clusters. Science 2008, 322, 1819–1822.
[61] Pouget EM, Bomans PHH, Goos JACM, Frederik PM, de With G, Sommerdijk NAJM. The initial stages of template-controlled $CaCO_3$ formation revealed by cryo-TEM. Science 2009, 323, 1455–1458.
[62] Cartwright JHE, Checa AG, Gale JD, Gebauer D, Sainz-Díaz CI. Calcium carbonate polyamorphism and its role in biomineralization: How many amorphous calcium carbonates are there? Angew Chem Int Edit 2012, 51, 11960–11970.
[63] Meyer H-J. Über Vaterit und seine Struktur. Angew Chem 1959, 71, 678–679.
[64] Kamhi SR. On the structure of vaterite, $CaCO_3$. Acta Crystallogr 1963, 16, 770–772.
[65] Meyer HJ. Struktur und Fehlordnung des Vaterits. Z Kristallogr 1969, 128, 183–212.
[66] Lippmann F. Sedimentary Carbonate Minerals. New York, NY, USA, Springer-Verlag, 1973.
[67] Le Bail A, Ouhenia S, Chateigner D. Microtwinning hypothesis for a more ordered vaterite model. Powder Diffr 2011, 26, 16–21.
[68] Lin J, Sheptyakov D, Wang Y, Allenspach P. Structures and phase transition of vaterite-type rare earth orthoborates: A neutron diffraction study. Chem Mater 2004, 16, 2418–2424.
[69] Wang J, Becker U. Structure and carbonate orientation of vaterite ($CaCO_3$). Am Mineral 2009, 94, 380–386.
[70] Demichelis R, Raiteri P, Gale JD, Dovesi R. A new structural model for disorder in vaterite from first-principles calculations. CrystEngComm 2012, 14, 44–47.
[71] Wang J, Becker U. Energetics and kinetics of carbonate orientational ordering in vaterite calcium carbonate. Am Mineral 2012, 97, 1427–1236.
[72] Demichelis R, Raiteri P, Gale JD, Dovesi R. The multiple structures of vaterite. Cryst Growth Des 2013, 13, 2247–2251.
[73] Wang J, Zhang F, Zhang J, Ewing RC, Becker U, Cai Z. Carbonate orientational order and superlattice structure in vaterite. J Cryst Growth 2014, 407, 78–86.
[74] De La Pierre M, Demichelis R, Wehrmeister U et al. Probing the multiple structures of vaterite through combined computational and experimental Raman spectroscopy. J Phys Chem C 2014, 118, 27493–27501.
[75] Kabalah-Amitai L, Mayzel B, Kauffmann Y et al. Vaterite crystals contain two interspersed crystal structures. Science 2013, 340, 454–457.
[76] Falini G, Fermani S, Reggi M, Njegić Džakula B, Kralj D. Evidence of structural variability among synthetic and biogenic vaterite. Chem Commun 2014, 50, 15370–15373.
[77] Dietzsch M, Andrusenko I, Branscheid R, Kolb U, Tremel W. Snapshots of calcium carbonate formation – A step by step analysis. Cryst Eng Comm 2015, submitted.
[78] Pasteris JD, Wopenka B, Valsami-Jones E. Bone and tooth mineralization: Why apatite? Elements 2008, 4, 97–104.
[79] Leroy G, Penel G, Leroy N., Brès E. Human tooth enamel: A Raman polarized approach. Appl Spectrosc 2002, 56, 1030–1034.
[80] Elliott JC. Monoclinic space group for hydroxyapatite. Nature Physical Science 1971, 230, 72.
[81] Elliott JC, Mackie PE, Young RA. Monoclinic hydroxyapatite. Science 1973, 180, 1055–1057.
[82] Busch S, Dolhaine H, DuChesne A et al. Biomimetic morphogenesis of fluorapatite-gelatin composites: Fractal growth, the question of intrinsic electric fields, core/shell assemblies, hollow spheres and reorganization of denatured collagen. Eur J Inorg Chem 1999, 1643–16453.
[83] Izu Y, Sun M, Zwolanek D et al. Type XII collagen regulates osteoblast polarity and communication during bone formation. J Cell Biol 2011, 193, 1115–1130.

[84] Lang SB, Tofail SAM, Gandhi AA et al. Pyroelectric, piezoelectric, and photoeffects in hydroxyapatite thin films on silicon. Appl Phys Lett 2011, 98, 123703.

[85] Palatinus L, Klementová M, Dřínek V, Jarošová M, Petříček V. An incommensurately modulated structure of η'-phase of $Cu_{3+x}Si$ determined by quantitative electron diffraction tomography. Inorg Chem 2011, 50, 3743–3751.

[86] Gemmi M, Campostrini I, Demartin F, Gorelik TE, Gramaccioli CM. Structure of the new mineral sarrabusite, $Pb_5CuCl_4(SeO_3)_4$, solved by manual electron-diffraction tomography. Acta Crystallogr B 2012, 68, 15–23.

[87] Zhang D, Oleynikov P, Hovmöller S, Zou X. Collecting 3D electron diffraction data by the rotation method. Z Kristallogr 2010, 225, 94–102.

[88] Yun Y, Zou X, Hovmöller S, Wan W. Three-dimensional electron diffraction as a complementary technique to powder X-ray diffraction for phase identification and structure solution of powders. IUCrJ 2015, 2, 267–282.

[89] Gemmi M, Oleynikov P. Scanning reciprocal space for solving unknown structures: energy filtered diffraction tomography and rotation diffraction tomography methods. Z Kristallogr 2013, 228, 51–58.

[90] Palatinus L, Jacob D, Cuvillier P, Klementová M, Sinkler W, Marks LD. Structure refinement from precession electron diffraction data. Acta Crystallogr A 2013, 69, 171–188.

[91] Palatinus L, Petříček V, Antunes Corrêa C. Structure refinement using precession electron diffraction tomography and dynamical diffraction: theory and implementation. Acta Crystallogr A 2015, 71, 235–244.

[92] Kaiser U, Biskupek J, Meyer JC et al. Transmission electron microscopy at 20-kV for imaging and spectroscopy. Ultramicroscopy 2011, 111, 1239–1246.

[93] de Jonge N, Peckys DB, Kremers GJ, Piston DW. Electron microscopy of whole cells in liquid with nanometer resolution. P Natl Acad Sci USA 2009, 106, 2159–2164.

[94] Nannenga BL, Shi D, Leslie AGW, Gonen T. High-resolution structure determination by continuous-rotation data collection in MicroED. Nat Methods 2014, 11, 927–930.

[95] Nederlof I, van Genderen E, Li Y-W, Abrahams JP. A Medipix quantum area detector allows rotation electron diffraction data collection from submicrometre three-dimensional protein crystals, Acta Crystallogr D 2013, 69, 1223–1230.

Frank Gfeller

7 Mayenite $Ca_{12}Al_{14}O_{32}[X^{2-}]$: From minerals to the first stable electride crystals

7.1 The discovery and chemical composition of mayenite $Ca_{12}Al_6Al_8O_{32}[O]$

The story of mayenite $Ca_{12}Al_{14}O_{32}[O]$ starts at the beginning of the 20th century. As early as 1909 a binary phase diagram between CaO and Al_2O_3, reporting four stable intermediate phases, had been developed [1]. Almost midway between the pure oxides, a pentacalcium-trialuminate $Ca_5Al_6O_{14}$ compound was reported as an optically isotropic phase with a melting point at approximately 1455 °C. The same phase formed together with CaO during the dehydration of 3 $CaO \cdot Al_2O_3 \cdot 6\ H_2O$ [2], a composition with a stoichiometry reminiscent of the hydrogarnet endmember katoite, $Ca_3Al_2(O_4H_4)_3$ [e.g. 3]. Although the stoichiometry of $Ca_5Al_6O_{14}$ is significantly different from $Ca_3Al_2(O_4H_4)_3$, the two X-ray diffraction patterns showed notable similarities suggesting structural relations. The structure of $Ca_3Al_2(O_4H_4)_3$ is of the garnet-type with vacancies at the tetrahedral site and hydroxylated $Al(OH)_6$ octahedra [3]. In an attempt to clarify the structural relation, Büssem and Eitel [4] synthesized and examined $Ca_5Al_6O_{14}$ using various photographic X-ray diffraction techniques combined with density measurements. Surprisingly, it was not possible to synthesize pure $Ca_5Al_6O_{14}$; the products always contained birefringent grains intermingled with the optically isotropic main phase. A further problem occurred when the number of formula units (Z) was calculated based on the cubic unit cell (point group $\bar{4}3m$, a = 11.95 Å) and the measured density. For the assumed composition $Ca_5Al_6O_{14}$ a Z of 4.735 was obtained, which seemed improbable since it was too far off an integer value. Furthermore, with $Z = 5$, the number of atoms in the cell would have been odd, which is incompatible with the considered cubic space groups. Based on these observations, Büssem and Eitel [4] concluded that the compound must have a molar ratio differing slightly from 5 CaO : 3 Al_2O_3 and calculated $Z = 2.00$ for the corresponding ratio of 12 : 7, which only required a small shift towards CaO in the binary system [4]. Subsequent synthesis attempts applying the 12 : 7 oxide ratio yielded a single phase cubic compound. The unit cell thus contained 24 Ca, 28 Al, and 66 O atoms. The next difficulty, however, arose with the assignment of space group $I\bar{4}3d$ No. 220. In this space group the available site multiplicities (12, 16, 24, 48) do not allow the accommodation of 66 O atoms in a structure. Nevertheless, a non-stoichiometric structure model with 64 O positions was calculated and consistent with the observed X-ray diffraction intensities. The two missing O atoms per unit cell were assumed to be statistically distributed within the structure. A distinct position was ruled out since no superstructure reflections had been observed. As the question about the position of the two additional O atoms remained unresolved, the suitability of the structure model [4] was accepted albeit with reservation. However, all subsequent studies confirmed its correctness [5–7].

7.2 The crystal structure of mayenite and its relation to the garnet structure

As indicated by the dehydration experiments on katoite $Ca_3Al_2(O_4H_4)_3$ [2] the crystal structure of mayenite is closely related to the garnet structure. While garnets crystallize in space group $Ia\bar{3}d$, mayenite belongs to the non-centrosymmetric subgroup $I\bar{4}3d$. The garnet structure is commonly described by the cation sites *X* (dodecahedral), *Y* (octahedral) and *Z* (tetrahedral). In the mayenite structure the site *T*1, which corresponds to *Y*, is tetrahedrally instead of octahedrally coordinated by O. The second tetrahedral site *T*2, corresponding to *Z* is only half occupied, resulting in the formation of structural cavities. These cavities are partially occupied by extra anions. The relation between katoite, mayenite and grossular is shown in Fig. 7.1, while the corresponding crystallographic sites and Wyckoff positions are listed in Tab. 7.1.

The mayenite structure can be described as a zeolite-like framework of AlO_4 tetrahedra. The *T*2 tetrahedra are connected to four *T*1 tetrahedra, while *T*1 shares only three O ligands with *T*2 tetrahedra. The fourth apex of the *T*1 tetrahedra (O2) points towards the structural cavities. All structural cages, about 5 Å in diameter, contain two Ca atoms (Fig. 7.2). In garnet, katoite and grossular, Ca is eight-fold coordinated, whereas in mayenite it is six-fold. In the latter, Ca has trigonal prismatic coordination with 2 ×Ca-O1 (2.38 Å), 2 ×O1 (2.49 Å), and 2 ×O2 (2.41 Å) (Fig. 7.2).

7.3 Extra framework anions and structure distortion

A first indication for the actual site of the extraframework oxygen was found in the course of the examination of the chloride-, hydroxide- and fluoride-derivates of mayenite [8]. In subsequent studies the crystal structure of $Ca_{12}Al_{14}O_{32}F_2$ has been refined [9, 10]. Fluorine occupies two of the six structural cages at their centre. The Coulomb force of F^- causes splitting of the adjacent Ca sites. In case of an F-occupied cage Ca is dragged by about 0.5 Å towards the cage centre to the new site Ca1a. This position is sevenfold coordinated with four normal Ca-O bonds (≈ 2.36 Å), two long bonds (≈ 2.94 Å), and one Ca-F bond (≈ 2.31 Å) [10].

Although the position at the centre of the cage was an appropriate first guess for the extraframework oxygen site (O3) in $Ca_{12}Al_{14}O_{32}[O]$, it took about forty years until this question was revisited as a scientific topic. Ab-initio calculations for several potential extraframework anions (OH^-, F^-, O^-, H^-, O^{2-}) suggested that small mono-

Tab. 7.1: Crystallographic sites and Wyckoff positions of katoite, mayenite, and grossular.

	Katoite $Ca_{12}{}^{VI}Al_8\square_{12}(OH)_{48}$	Mayenite $Ca_{12}{}^{IV}Al_8{}^{IV}Al_6\square_6^* O_{32}$ [O]*	Grossular $Ca_{12}{}^{VI}Al_8{}^{IV}Si_{12}O_{48}$
Space Group	$Ia\bar{3}d$ (230)	$I\bar{4}3d$ (220)	$Ia\bar{3}d$ (230)
X	Ca 0, $\frac{1}{4}$, $\frac{1}{8}$ (24 *c*)	Ca x, 0, $\frac{1}{4}$ (24 *d*) $x \approx 0.13$	Ca 0, $\frac{1}{4}$, $\frac{1}{8}$ (24 *c*)
Y	Al 0, 0, 0 (16 *a*)	Al *x*, *x*, *x* (16 *c*) $x \approx 0.18$	Al 0, 0, 0 (16 *a*)
Z	□ $\frac{3}{8}$, 0, $\frac{1}{4}$	Al $\frac{3}{8}$, 0, $\frac{1}{4}$ □* $\frac{7}{8}$, 0, $\frac{1}{4}$ (12 b)*	Si $\frac{3}{8}$, 0, $\frac{1}{4}$
	OH *x*, *y*, *z* (96 *h*)**	O1 *x*, *y*, *z* (48 *e*) O2 *x*, *x*, *x* (16 *c*)**	O *x*, *y*, *z* (96 *h*)**

*The 33rd extra frame work oxygen is located in one of the 6 structural cages slightly off the cage centre ($\frac{7}{8}$, 0, $\frac{1}{4}$) at a general position *x*, *y*, *z*.
**The octahedral site *Y* in garnet corresponds to the tetrahedral site *T*1 in mayenite. Half of the O-sites remain at a general position (48 *e*) and form the base triangle of the *T*1 tetrahedron (O1). The remaining 48 O of the garnet structure are replaced by 16 (O2) in the mayenite structure and form the apex of the *T*1 tetrahedron (O2).

valent anions are located at the cage centre, while the bivalent O^{2-} is shifted towards the cage wall, strongly interacting with the adjacent Ca site and Al1 [11]. Two detailed experimental studies on the local structure of occupied cages in mayenite were published almost simultaneously in 2007: (1) A neutron powder-diffraction study at temperatures between RT and 1323 K [12] and (2) joint Rietveld refinements of synchrotron X-ray and neutron data combined with single-crystal X-ray diffraction data [13]. Both studies agree on a split site for Ca depending on the presence of an extraframework anion, while the tetrahedral framework is reported to remain unaffected, as found for $Ca_{12}Al_{14}O_{32}F_2$ [10, 12, 13]. The residual electron density within the structural cages has been discussed in detail. At room temperature, Boysen et al. [12] observed the main electron density at the centre of the cage. At elevated temperature the residual peaks gradually shifted towards the cage wall resulting in four symmetrically equivalent peaks arranged around the $\bar{4}$-axis at a distance of about 0.8 Å from the centre. The refined occupancy factor for the extraframework O3, which was placed at the cage centre, exceeded the theoretical values for stoichiometric $Ca_{12}Al_{14}O_{32}$[O], particularly at lower temperature. This excess population was assigned to the incorporation of OH^- (due to subsequent partial hydroxylation of O^{2-} to OH^-, see discussion below).

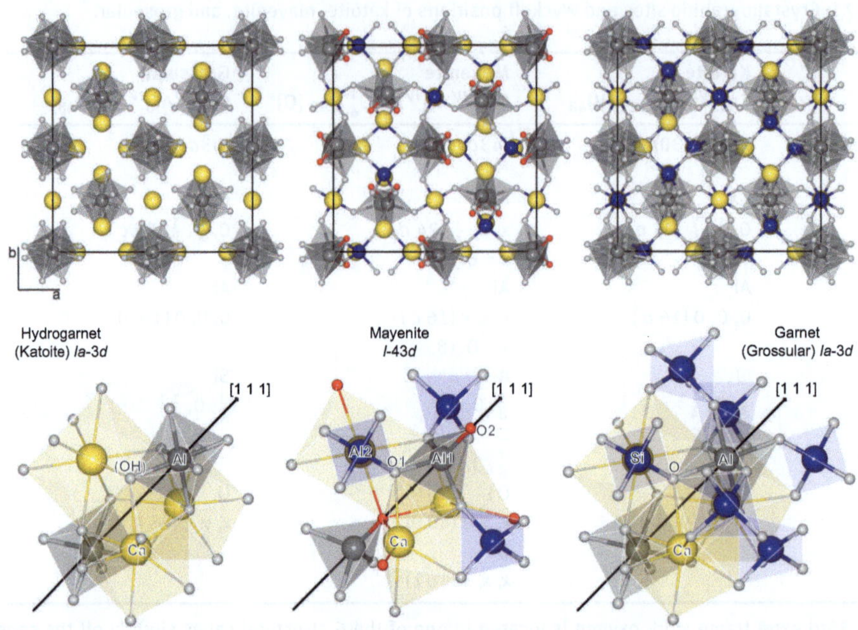

Fig. 7.1: Structural comparison of katoite (left), mayenite (centre), and grossular (right). The main difference between garnet and mayenite is that garnet has octahedral Al whereas in mayenite the Al1 site is tetrahedrally coordinated (dark grey). Three oxygen atoms (light grey) in garnet are replaced by one O2 site (red) in the mayenite structure. The second tetrahedrally coordinated site (Al2, blue) in mayenite corresponds to half of the Si sites in grossular (dark blue). The position of the second half of tetrahedra in the grossular structure is occupied by cavities in mayenite. In katoite this site remains vacant.

The residual density off the centre has been attributed to an O_2^- radical. The results of Boysen et al. [12] are generally in agreement with Palacios et al. [13]. In a subsequent single-crystal X-ray study, however, the latter [14] propose a more precise model. Beside the O position in the cage centre, which still hosts the main electron density, four symmetrically equivalent sites (O3) around the $\bar{4}$ axis are reported. In addition, a split position for Al1 (Al1a) was included in the model. The distance between the newly defined Al1a site and O3 corresponds to an appropriate Al-O bond length of 1.73 Å. This model indicates two possible extraframework O sites, one at the centre and one slightly off [14]. In contrast to the previous results [12–14] a more recent synchrotron single-crystal X-ray study reports only an off-centre displaced extraframework O, occupying one of four symmetrically equivalent sites around the $\bar{4}$ axis [15]. In addition, a detailed model of the local structure distortion including several split positions for all sites except Al2 has been developed. In contrast to previous studies all occupancies

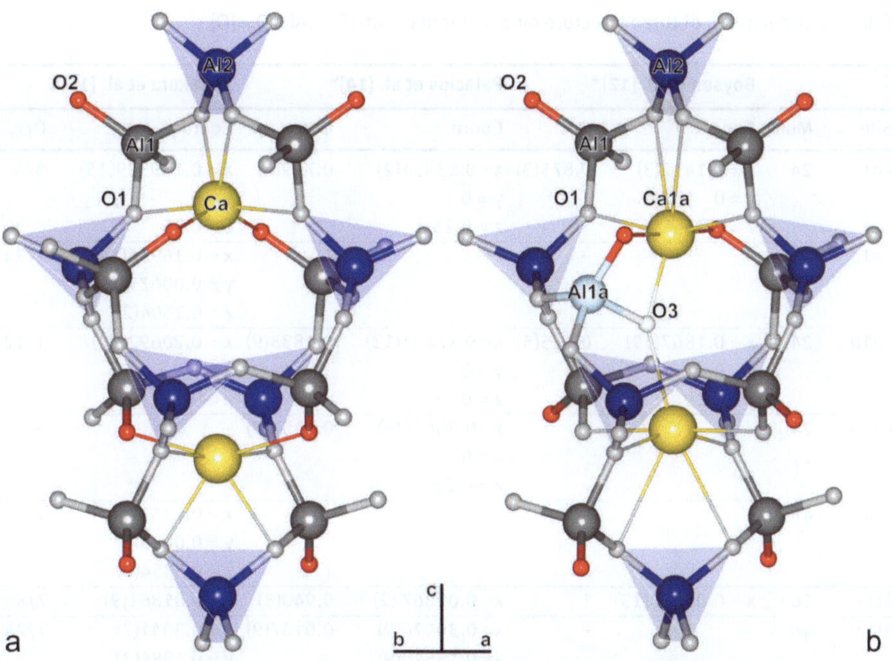

Fig. 7.2: (a) A single cage of the mayenite structure without potential O3: each cage, about 5 Å in diameter, is filled by two Ca atoms (yellow spheres) (b) The presence of extra-framework oxygen (O3) causes the shift of the two Ca atoms towards the cage centre. Moreover, O3 forms a new bond Al1a-O3 with the adjacent Al1a site while another Al1-O1 bond is broken. Al2 tetrahedra are shaded blue whereas Al1 tetrahedra are presented only by ball and sticks. O2 is in red, O1 in gray and O3 in white.

of split positions were fixed to theoretical values to avoid correlation problems during structure refinement. A comparison of the various structure models is given in Tab. 7.2.

7.3.1 Centre, off-centre or both?

After careful consideration of the hitherto published structure models, the exact nature of the extraframework O site still remains ambiguous. Either O3 is located at the cage centre [12, 13], or slightly displaced from the centre forming a new Al-O bond with the split site Al1a [15]. Both cases could co-exist simultaneously [14].

A hypothesis for the cause of the additional electron densities in the cage centre is the hydration or hydroxylation of mayenite. It is broadly discussed that mayenite is a hygroscopic material, which may even hydrate at normal humidity [8, 16–18]. H_2O reacts with the cage filling O^{2-} according to $H_2O_{(g)} + O^{2-}_{(cage)} \rightarrow 2OH^{-}_{(cage)}$ [16, 17]. Especially powders having larger specific surfaces than single crystals may easily hydrate

Tab. 7.2: Comparison of three structure models for mayenite $Ca_{12}Al_{14}O_{32}[O]$

		Boysen et al. [12]*		**Palacios et al. [14]***		**Sakakura et al. [15]***	
Site	Mult.	Coord.	Occ.	Coord.	Occ.	Coord.	Occ.
Ca1	24	x = 0.1432(3) y = 0 z = 0.25	0.875(3)	x= 0.13831(2) y = 0 z = 0.25	0.769(2)	x = 0.139339(15) y = 0 z = 0.25	3/4
Ca1a	48	–	–	–	–	x = 0.16936(16) y = 0.00673(17) z = 0.2504(2)	1/24
Ca1b	24	x = 0.1867(19) y = 0 z = 0.25	0.125(3)	x= 0.17451(13) y = 0 z = 0.25	0.1833(9)	x = 0.20692(12) y = 0 z = 0.25	1/12
Ca1b*	24	–	–	x= 0.20650(4) y = 0 z = 0.25	0.0457(8)	–	–
Ca1c	48	–	–	–	–	x = 0.13537(17) y = 0.0043(3) z = 0.2354(2)	1/24
Al1	16	x = 0.01880(13)	1	x= 0.01867 (2)	0.940(3)	x = 0.01861(9)	7/8
Al1a	48	–	–	x= 0.30520(9) y= 0.19510(9) z= 0.26090(9)	0.0167(9)	x = 0.3041(2) y = 0.1984(2) z = 0.2619(2)	1/24
Al2	12	x = 0.875 y = 0 z = 0.25	1	x = 0.875 y = 0 z = 0.25	1	x = 0.875 y = 0 z = 0.25	1
O1	48	x = 0.03672(10) y = 0.44292(10) z = 0.15054(9)	1	0.03612(4) 0.44231(4) 0.15049(4)	1	x = 0.03605(4) y = 0.44234(3) z = 0.15052(3)	11/12
O1a	48	–	–	–	–	x = 0.1302(3) y = 0.0603(3) z = 0.43286(28)	1/24
O1b	48	–	–	–	–	x = 0.1718(5) y = 0.0312(4) z = 0.4352(5)	
O2	16	x= 0.18519(9)	1	0.18526(3)	1	x= 0.18510(4)	7/8
O2a	48	–	–	–	–	x = 0.1996(6) y = 0.1750(7) z = 0.1893(6)	1/24
$O3_{(Cen.)}$	12	x= 0.375 y = 0 z = 0.25	0.251(6)	x= 0.375 y = 0 z = 0.25	0.106(4)	–	–
$O3_{(off)}$	48	–	–	0.3588(2) 0.0616(2) 0.2479(2)	0.018(2)	x = 0.3559 (6) y = 0.0614 (7) z = 0.2506 (7)	1/24

* Coordinates from the original publications have been changed to symmetry equivalent positions where necessary to ensure comparability.

under ambient conditions after sample preparation. In order to prove this hypothesis, we synthesized single crystals of pure $Ca_{12}Al_{14}O_{33}$ from a melt in platinum crucibles. $CaCO_3$ (Sigma-Aldrich 239216) and Al_2O_3 (Sigma-Aldrich 23474-5) were mixed in a molar ratio of 12 : 7 and heated up to 1500 °C. Subsequently the melt was cooled to 1200 °C with a cooling rate of 20 °C/h. The product is an aggregate of mayenite single-crystals. A first series of single-crystals was separated and immediately measured on a Bruker APEXII diffractometer. A second batch was exposed to a humid atmosphere for several days. The Crystallographic Information Files (CIF) including details about the X-ray measurements have been submitted to ICSD FIZ Karlsruhe.

7.3.2 Extraframework occupants in "dry" mayenite

The crystal-structure refinement of the dry samples has been performed in a stepwise procedure starting with the tetrahedral framework and Ca position. After this first refinement $R1 = \Sigma\|Fo| - |Fc\|/\Sigma|Fo|$ reduced to 0.0765. The highest remaining peak (3.29 e^-) in the difference-Fourier map is due to the split Ca position (Fig. 7.3 a). The next procedure involved the definition of a refined variable parameter X, which may be regarded as the ratio between empty and occupied cages (empty = 12 X, occupied = 12(1 – X)). If the site multiplicity is taken into account, the site occupancy factors (*sof*) for the two split Ca sub-positions are defined as $sof_{Ca1} = 0.5X$ and $sof_{Ca1a} = 0.5(1-X)$. After this step, *R*1 drops to 0.0362 with a residual peak of 0.65 e^- close to Al1 due to the splitting of the Al1 site (Fig. 7.3 a). The *sof* for the split position (Al1a) has been refined as $sof_{Al1} = 1/3X$ and $sof_{Al1a} = 1/3(1-X)$. As a result of this third step *R*1 drops further to 0.019 while the highest residual peak (0.42 e^-) remains close to the cage centre and is caused by the extra framework O^{2-} (O3) (Fig. 7.3 b). The position is split in to 4 sub-sites around the $\bar{4}$ axis and the *sof* has been set as $sof_{O3} = 0.25(1-X)$. The final refinement converged at *R*1 of 0.0113 with a residual maximum of 0.18 e^- close to Ca1 and a deepest hole of 0.28 e^-.

Our refinement of $Ca_{12}Al_{14}O_{32}[O]$ confirms the model with the off-centre extraframework oxygen position [15]. In contrast to Sakakura et al. [15], where the occupancies of each split position have been fixed, we refine one single variable X for all positions. In the structure of mayenite, $Ca_{12}Al_{14}O_{32}[O]$, one of six cages is occupied by O^{2-} and the expected value for X becomes 0.8333. In contrast, in our refinement X converges to 0.889(2) yielding a sum formula of $Ca_{12}Al_{14}O_{32}[O_{0.66(2)}]$. An explanation for this discrepancy is given by the simplification of the Ca-splitting. Since the extraframework O is not centred in the cage, two additional Ca-positions (Ca1a, Ca1b) would be a more accurate model. However, the refinement of such a model is strongly affected by correlations and does not further contribute to the understanding of the extraframework O position.

The new data confirms [15] that the local presence of O3 causes breaking of an Al1-O1 bond associated with the formation of a new Al1a-O3 bond. As a consequence, the

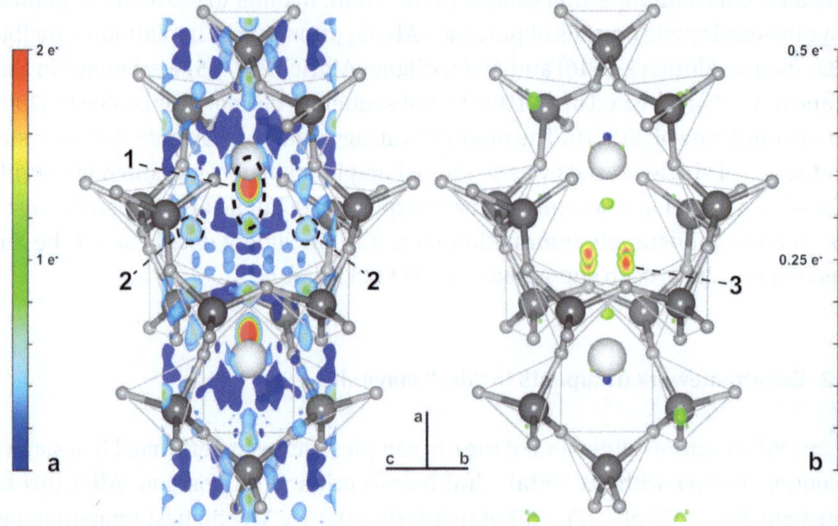

Fig. 7.3: Two difference Fourier-maps (010) and (001) are plotted within the structure model of an empty cage. Residual densities <0.2 e^- are not plotted. (a) The difference Fourier map of dry mayenite after refinement of the tetrahedral framework plus Ca sites. The highest peak (1) is caused by the splitting of Ca. Furthermore, residual density (2) close to Al1 indicates the split site Al1a. (b) After the refinement of the split positions for Al1 and Ca1 the remaining maximum is caused by O3 located slightly off the centre of the cage (3).

mayenite structure contains not only one loosely bonded O site (O3) in the cage, but also an additional O1 site, which is connected once to Al2 and twice to Ca. Moreover, Al1a forms a new tetrahedron with O3 as one of the apices. Therefore, the question arises whether the term extraframework oxygen is appropriate for O3.

7.3.3 Hydroxylated mayenite

In contrast to the measurements of dry mayenite, the samples exposed to humidity show a pronounced residual electron density in the centre of the cage caused by OH^- (Fig. 7.4) similar to the previously reported Fourier maps [12–14]. Since the O^{2-}/OH^- ratio of hydroxylated mayenite is not known, a refinement with theoretical occupancies is not feasible. To solve this issue two refined variables have been defined: (1) X based on the splitting of Ca1 and Ca1a similar to X in dry mayenite and (2) Y based on the splitting of Al1 and Al1a, ($sof_{\text{Al1}} = 1/3\text{Y}$, $sof_{\text{Al1a}} = 1/3(1-\text{Y})$). The refined value for X converges to 0.75(2) that is significantly lower than in dry mayenite and corresponds to 25(2) % cages occupied by O^{2-} and OH^-. The occupancy of the displaced O3 has been coupled to Y ($sof_{O3} = 0.25(1-\text{Y})$) since O3 causes the splitting of Al1. Y converges to 0.924 (2) indicating that 7.57(3) % of the cages are occupied by O^{2-}. Assuming Ca1a to be occupied only in case of O3, the occupancy of OH^- at the cage centre has been calculated and fixed according to $sof._{OH-} = 0.25(Y-X)$ resulting in 17 % of the cages occupied by OH^-. The refinement converges to $R1 = 0.013$ with maximum residuals of 0.21 e^- and −0.33 e^-. The charge balance of the resulting sum formula $Ca_{12}Al_{14}O_{32}[O_{0.45(1)}OH_{1.02}]$ amounts to +66 cationic charges / -65.92 anionic charges and is a further indicator for the validity of the refined model.

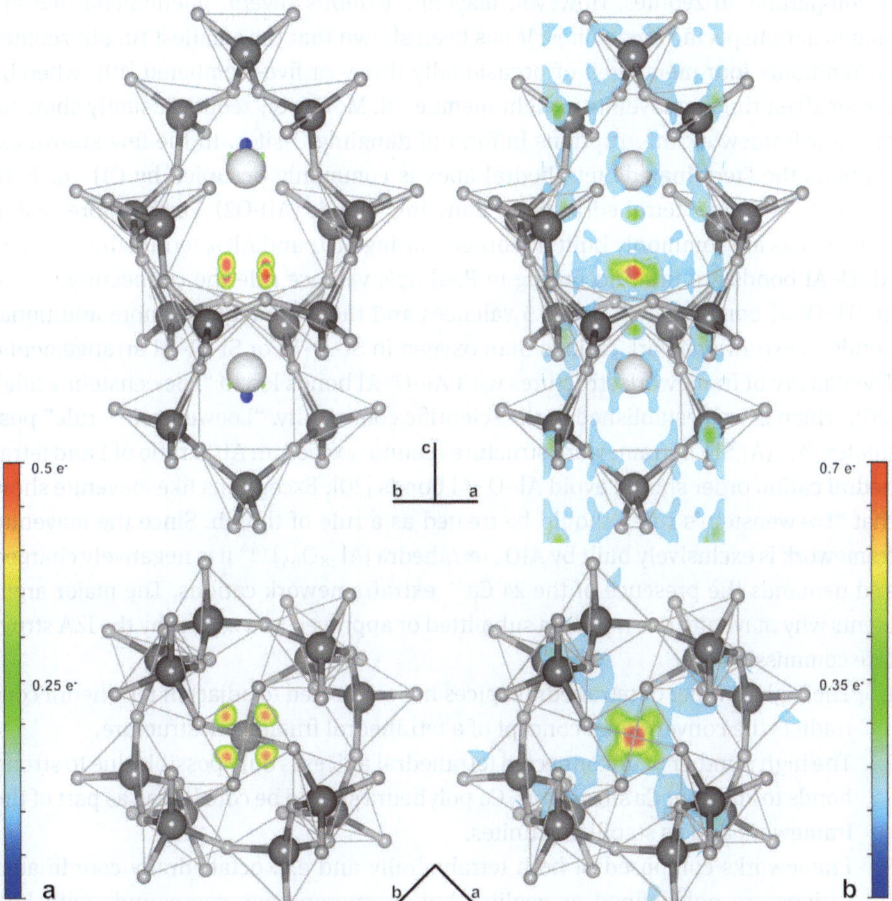

Fig. 7.4: A comparison of difference Fourier-maps of dry (a) and hydroxylated (b) mayenite along [110] (top) and [001] (below). After refinement of the model with split positions for Ca and Al, residual peaks indicate the position of the extraframework anions. In (a) only O3 is at x,y,z slightly displaced from the cage-centre. In (b) additional density is observed at the centre at 0.375, 0, 0.25 indicating the presence of OH^-. Residuals < 0.2 e^- have not been plotted.

7.4 Zeolite or anti-zeolite?

Due to its tetrahedral framework and ion exchange capacity mayenite is often described as a zeolite-like compound. Therefore the question arises, whether mayenite may be regarded as zeolite. There is no definite criterion, which excludes mayenite from an inclusion to the atlas of zeolite structure-types. Zeolites are microporous compounds built by a tetrahedral framework with framework densities (tetrahedral sites per 1000 $Å^3$) between 12.5 and 20.2 [19]. The framework density of mayenite ≈ 16.28 is comparable to zeolites. However, mayenite exhibits several phenomena that are somewhat unspecific for zeolites. It has been shown that the smallest ring in zeolites is commonly four-membered or occasionally three- or five-membered [19], whereby the smallest ring in mayenite is eight-membered. Moreover, zeolites usually show no or a few framework interruptions in form of dangling O-sites. In the few known exceptions the "terminated" tetrahedral apex is commonly occupied by OH^- or F. In mayenite 4 out of 7 tetrahedra show "dangling" bonds (Al1-O2). Furthermore zeolite frameworks are commonly built by corner-sharing SiO_4 and AlO_4 tetrahedra whereby Al–O–Al bonds are rare. According to Pauling's valance rule the connecting O^{2-} in an Al–O–Al bond receives only 1.5 valences and therefore requires more additional bonds to extraframework cations than oxygen in Si–O–Si or Si–O–Al arrangements. The paucity of framework structures with Al–O–Al bonds led to "Loewenstein's rule" [20], which is well established in the scientific community. "Loewenstein's rule" postulates that $(Al,Si)O_4$-framework structures cannot exceed an Al/Si ratio of 1 and tetrahedral cation order should avoid Al–O–Al bonds [20]. Exceptions like mayenite show that "Loewenstein's rule" should be treated as a rule of thumb. Since the mayenite framework is exclusively built by AlO_4 tetrahedra $[Al_{28}O_{64}]^{-44}$ it is negatively charged and demands the presence of the 24 Ca^{2+} extraframework cations. The major arguments why mayenite has not been submitted or approved as a zeolite by the IZA structure commission are:

1. The high number of tetrahedral apices not connected to adjacent tetrahedra contradicts the conventional concept of a tetrahedral framework structure.
2. The high number of unconnected tetrahedral apices is only possible due to strong bonds to adjacent Ca sites. Thus, Ca polyhedra should be considered as part of the framework and as stabilizing unites.
3. Frameworks composed of both tetrahedrally and e.g. octahedrally coordinated cations are not defined as zeolites but as microporous compounds with heteropolyhedral framework structures.

Recently, authors [14, 15, 21] classified mayenite as "anti-zeolite" due to its anion exchange capacity. Regarding all similarities between zeolites and mayenite the term "anti-zeolite" is rather misleading. In conclusion, mayenite is neither a zeolite nor is the term anti-zeolite appropriate to describe it.

7.5 Applications of mayenite ceramics

Due to its microporous structure, anion-exchange-capacity and ion-conductivity, mayenite may be used for a variety of applications such as a transparent oxide conductor (TOC) [22], electron emission source [23], catalyst [24, 25], and O^- ion source [23]. A remarkable strength of mayenite as an advanced material is the availability of the constitutional elements Ca, Al, O and the low production cost of the synthesis. In the following an overview of mayenite applications is presented.

7.5.1 Mayenite the first thermally and chemically stable electride

Electrides are a rare class of compounds in which electrons behave as anions. Usually electrons are confined to atoms, ions or molecules. In metals, they are delocalized in an "electron gas". Electrides however, incorporate electrons, which replace anion sites usually occupied by Cl^-, F^-, OH^-, within a crystal lattice [26]. Such electron sites were first reported as lattice defects in alkali salts such as KCl [27] and are known as *F*-centres. In crystalline electrides, anion sites are stoichiometrically occupied by *F*-centres [26]. The first reported electrides were organic alkali-electrides in which alkali ions are bonded to nonreducible complexants such as crown ethers. The electrons act as anions occupying voids in between these complexes [26]. Examples of such electrides are Li^+(cryptand [2.1.1])e^- [28], K^+(cryptand [2.2.2])e^- [29], $Cs^+(18C6)_2 \cdot e^-$ [30] etc. Although it is possible to crystallize these compounds, most known organic electrides are very sensitive to air, moisture and decompose at temperatures above −40 °C [26]. A strategy to overcome this problem is the synthesis of inorganic electrides. Electrons may be incorporated in channels of microporous aluminosilicates (zeolites) by addition of vaporized alkali metals [31]. If electrons are added to the channels in zeolites, they do not remain free but build clusters with the extraframework cations. A well investigated example is electron-doped sodalite, known as "black sodalite" where electrons form clusters of Na_4^{3+} in structural cages [e.g. 32]. For a successful synthesis of a zeolite-based electride the separation of channels or cages containing extraframework cations and others with entrapped electrons would be a requisite [26].

Electron entrapment in mayenite was firstly realized by the incorporation of H^- ions in structural cages by thermal treatment in a hydrogen atmosphere [33]. The resulting hydrogen-bearing mayenite is an insulator but may be converted in to an electronic conductor upon illumination with UV-light. The formation of electrons in the cages is reported according to the reaction $H^-_{(cage)} + O^-_{2(cage)} + vac._{(cage)} \leftrightarrow 2e^-_{(cage)}OH^-_{(cage)}$ [23]. The material changed colour from colourless transparent to yellowish green translucent and revealed a conductivity of 0.3 S cm^{-1} at room temperature [33]. This photo-induced conversion from insulator to conductor drove attention as a potential new, economically favourable transparent oxide-conductor (TOC) material [22]. TOC's are used in solar-cells, defrosting windows, electrochromic windows and mirrors, touch-panel screens, transparent front electrodes in flat-panel displays, and many more applications [22]. However, the rather low conductivity of e^- doped mayenite compared to common TOC's demands further research [22].

A method to enhance the conductivity is the increase of carrier-doping. In the case of mayenite this is the entrapment of electrons in the structural cages whereby the maximum $Ca_{12}Al_{14}O_{32}[2e^-]$ has been successfully synthesized [34]. Electron doped mayenite $Ca_{12}Al_{14}O_{32}[2e^-]$ is the first stoichiometric, chemically and thermally stable electride and was produced by heat treatment of a mayenite single-crystal together with Ca metal-vapour in an evacuated silica-glass container. An electron concentration of 2×10^{21} e^- cm^{-3} has been reported, which is close to the theoretical maximum of 2.32×10^{21} e^- cm^{-3} [34]. The electrons are localized in the structural cages forming *F*-centres and are able to migrate to neighbouring cages. This electron mobility causes a conductivity of up to 100 S cm^{-1} [34].

Mayenite electrides may be produced in large quantities via reduced melt in sealed carbon crucibles at temperatures of approximately 1600 °C [35]. Moreover, electrons may be entrapped in mayenite using vanadium foil as a deoxygenating agent [13], or thermal treatment in Ti-metal vapour [36].

Due to its very low work function (minimum energy required to transfer an electron from the crystal to a point in the vacuum immediately outside the crystal surface), mayenite electrides show potential for future applications as electron emitters, e.g. in field-emission displays (FED). Moreover, mayenite electrides are thermally stable up to high temperature and may thus be used as thermal field emitters (TFE). Examinations on the thermal field emission of mayenite electrides between 0–6 kV acceleration voltage and temperatures from 600 °C up to 900 °C show that mayenite is a qualified material for TFE applications [23]. The emission with a current of $\approx$ 50 µA from an 80 μm^2 surface has been reported to stay stable for 90 h [23]. The mayenite electride may be regarded as the electron-emission source with the lowest work function [23].

7.5.2 Active oxygen O^-, O_2^- in mayenite as support of syngas catalysis

The need for new energy sources increases the importance of biomass as a feedstock. Partial oxidation of methane CH_4 into a mixture of H_2 and CO, known as syngas, is a modern route of fuel production [e.g. 37]. For the catalysis of this reaction, transition and noble metals supported by inert oxides such as Ni/MgO or Pt/MgO are common [e.g. 38]. Although the supporting oxides reduce the deposition of coke on the metals, the search for more efficient support materials is a current issue [24]. Active oxygen species, namely O^-, O_2^- and O_3^- play an essential role in the oxidation of hydrocarbons [39–41] and may be incorporated up to a considerable amount in mayenite [42]. Thus, mayenite is an interesting candidate as an "active" support of metal catalysts [24]. The first report on the presence of active oxygen species in mayenite goes back to Hayashi et al. [42]. Extraframework O^{2-} reacts with O_2 according to $O^{2-}_{(cage)} + O_{2(gas)} \rightarrow$

$O^-_{(cage)} + O^-_{2\text{-}(cage)}$. This reaction may be enhanced by a dry controlled atmosphere and elevated temperature. Annealing of mayenite in a hot isostatic pressing furnace under high oxygen fugacity allows concentrations of active oxygen species of up to 1.7 $\times 10^{21}$ cm^{-3} [43]. Syngas production has been studied over mayenite promoted with Ni, Co, Pt, Pd, and Ru, of which Ni shows promising results [24, 25]. Moreover, mayenite has been reported as a catalyst for the pyrolysis of *n*-hexane [44], *n*-heptane [45], methyl cyclohexane [46] as well as the selective N_2O removal from process gas of nitric acid plants [47].

Beside the favourable properties for catalysis, active oxygen species may be directly extracted from mayenite [23]. Selective extraction of O^- as a dense pure ion-beam was realized by applying DC voltage to an O_2 -treated mayenite sample at a working temperature of 800 °C and a voltage greater than 1 kV/cm [23, 48]. The measured O^- current ($\approx$ 2 μA/cm^2) exceeded the maximum value of yttria-stabilized zirconia (YSZ). Mayenite as source of a high density O^- ion beam may be applied for the decomposition of pollutants, sterilization, gate dielectrics on semiconductors, and for chemical reactions [23].

7.5.3 Cation and anion doping of mayenite

The applications summarized here triggered researchers to synthesize novel compounds with mayenite-type structure. An overview of the currently known mayenite related compounds is given by Schmidt [49]. In addition, the approach of mayenite doping with a variety of cations (Mg, Ga, Fe, Co, Ni, Cu, Zn, V, Nb, Ta, Y, In, Sc, Cr, Mn, Sn, Zr, Mo La, Ce, Pr, Nd, Sm, Eu, Gd, Tb, Dy, Ho, Er, Tm, Yb, Lu) is summarized [49]. Doping of mayenite with Si is currently the most discussed process. Based on the dehydration of hydrogarnet $Ca_3Al_2(SiO_4)_{3-x}(OH)_{4x}$ with $0 < x \leq 3$, compounds of the type $Ca_{12}Al_{14-2x}Si_{2x}O_{32}[O_{1+x}]$ may be synthesized [50]. The increased excess of positive charge (framework + Ca) allows a higher degree of active oxygen incorporation [50] and may lead to an enhanced conductor [51] compared to common mayenite. However, the synthesis of these compounds is far more demanding than that of common mayenite and the products are reported to be metastable [50, 51].

7.6 The mineral mayenite

The Ettringer Bellerberg Volcano near the city of Mayen is famous for a huge variety of rare minerals, formed due to metasomatic interaction of Ca-xenoliths with leucite-tephrite lava [52, 53]. The mineral mayenite was first reported from one of these Ca-rich xenoliths and named after the town of Mayen [54]. In analogy to the previously known compound $Ca_{12}Al_{14}O_{33}$ [4], the composition of mayenite was reported to be $Ca_{12}Al_{14}O_{33}$. However, the measured wet chemical composition remained unsatisfactory due to admixed impurities [54]. In a subsequent study, Hentschel, who originally described $Ca_{12}Al_{14}O_{33}$ as a mineral, reported mayenite to contain several wt.% Cl and suggested mayenite to be a solid-solution member between $Ca_{12}Al_{14}O_{32}[\square_5O]$ and $Ca_{12}Al_{14}O_{32}[\square_4Cl_2]$ [55]. In addition to the type locality, only two further terrestrial provenances are known. Mayenite was reported from the pyro-metamorphic rocks of the Hatrurim formation in Israel [56], as well in altered xenoliths in alkali basalts from Klöch, Styria [57]. Moreover, the Cl-rich variety of mayenite has been detected in a refractory inclusion of the Northwest Africa 1934 meteorite and defined as the mineral brearleyite [58].

While synthetic mayenite was extensively discussed in material science, the mineral mayenite stayed only scarcely examined until recently. In 2012 a re-investigation of mayenite from the type-locality indicated that the holotype specimen of mayenite (Fig. 7.5) is closer to the Cl-variety $Ca_{12}Al_{14}O_{32}[\square_4Cl_2]$ as well as hydroxylated [59]. Subsequent studies have shown that mayenite from the Hatrurim Formation is not pure either, but with a composition close to $Ca_{12}Al_{14}O_{32}F_2$ and additionally hydroxylated [60]. Furthermore, H_2O bearing varieties of these two endmembers were found: $Ca_{12}Al_{14}O_{32}[Cl_2(H_2O)_4]$ [61] and $Ca_{12}Al_{14}O_{32}[F_2(H_2O)_4]$ [60]. The most astonishing finding, from a crystallographic point of view, is the hydroxylation of natural mayenite. Beside the known OH^- position at the centre of the cages, three OH^- groups replace one O2-site, changing the coordination of an Al1 site from tetrahedral to octahedral (Fig. 7.6) [59, 60]. Neither this hydroxylation mechanism, $O^{2-}_{(O2)} + (F^-/Cl^-)_{(cage)} \rightarrow 3(OH^-)_{(O2a)}$, nor the incorporation of H_2O molecules have been reported for synthetic mayenite. Since $Ca_{12}Al_{14}O_{32}[\square_5O]$ is not the correct formula for type-locality mayenite, it has been redefined as chlormayenite $Ca_{12}Al_{14}O_{32}[\square_4Cl_2]$, which is identical with brearleyite. Subsequently the mineral name brearleyite was discredited, while the name mayenite stays reserved for compounds and a potential "new" mineral with endmember formula $Ca_{12}Al_{14}O_{32}[\square_5O]$ [61]. The new findings of mayenite-group minerals, including the first description and naming of the four minerals fluormayenite ($Ca_{12}Al_{14}O_{32}F_2$), fluorkyuygenite ($Ca_{12}Al_{14}O_{32}[F_2(H_2O)_4]$), chlorkyuygenite ($Ca_{12}Al_{14}O_{32}[Cl_2(H_2O)_4]$), and the isostructural silicate eltyubyuite ($Ca_{12}Fe_{10}Si_4O_{32}Cl_6$), gave rise to the introduction of a the new nomenclature scheme and definition of the mayenite supergroup [61].

7.6.1 The mayenite supergroup

The mayenite supergroup comprises all minerals isostructural with mayenite, subdivided in two subgroups: (1) mayenites (oxides) and (2) wadalites (silicates). The classification is based on the general formula $X_{12}T_{14}O_{32-x}(OH_{3x})[W_{6-3x}]$ with $0 \leq x \leq 2$. X refers to the Ca site, T to both tetrahedral sites and $[W]$ to the anions site in the structural cages. For the nomenclature the two tetrahedral sites $T1$ and $T2$ are considered together since many elements may substitute at both T-sites. The criterion for

Fig. 7.5: Holotype specimen of chlormayenite from an altered Ca-xenoltih of the Ettringer-Bellerberg volcano, (M5026/86 from the Mineral Museum, University of Cologne). The greenish zone is enriched in chlormayenite = (may) associated with larnite (β-Ca_2SiO_4) = (lar) and brownmillerite ($Ca_2(Al,Fe)_2O_5$) = (brm).

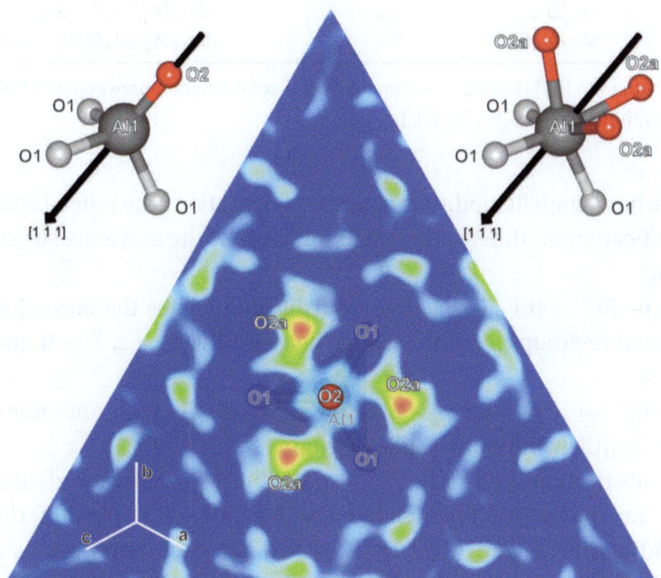

Fig. 7.6: Difference Fourier-map (111) after the refinement of the structure of fluormayenite from Hatrurim pyrometamorphic complex. Three residual peaks indicate a partial hydroxylation according to $O^{2-}_{(O2)} + F^{-}_{([W])} \rightarrow 3\times(OH^{-})_{(O2a)}$. This substitution changes the coordination of Al1 from tetrahedral to octahedral.

the subgroups is the sum of excess positive charges (framework + Ca) balanced by anions within the structural cavities at the [*W*] site. Endmembers of the mayenite group have a charge of 2– distributed over 6 [*W*] sites, while wadalites show 6– charges per 6 [*W*]-sites. The boundary between wadalites and mayenites is an excess charge of 4–. A list of all known minerals with mayenite type structure is given in Tab. 7.3. A third potential mineral group would be defined by the endmember $Ca_{12}Al_{14}O_{30}(OH_6)[\square_6]$ where two O2 sites are replaced by $2 \times (3OH^-)$ at O2a leading to a sum of charges at [*W*] of 0.

Tab. 7.3: Minerals of the mayenite supergroup

Mayenite supergroup	T site	W site	Formula
Mayenite group (W_{charge} = -2)			
1. chlormayenite	Al_{14}	$\square_4Cl_2$	$Ca_{12}Al_{14}O_{32}[\square_4Cl_2]$
2. fluormayenite	Al_{14}	$\square_4F_2$	$Ca_{12}Al_{14}O_{32}[\square_4F_2]$
3. chlorkyuygenite	Al_{14}	$(H_2O)_4Cl_2$	$Ca_{12}Al_{14}O_{32}[(H_2O)_4Cl_2]$
4. fluorkyuygenite	Al_{14}	$(H_2O)_4F_2$	$Ca_{12}Al_{14}O_{32}[(H_2O)_4F_2]$
Wadalite group (W_{charge} = -6)			
5. wadalite	$Al_{10}Si_4$	Cl_6	$Ca_{12}Al_{10}Si_4O_{32}[Cl_6]$
6. eltyubyuite	$Fe^{3+}_{10}Si_4$	Cl_6	$Ca_{12}Fe^{3+}_{10}Si_4O_{32}[Cl_6]$
7. adrianite*	$Mg_3Al_4Si_7$	Cl_6	$Ca_{12}Mg_3Al_4Si_7O_{32}[Cl_6]$*

*The reported formula for adrianite [62] is not an endmember but a solid solution between wadalite $Ca_{12}Al_{10}Si_4O_{32}[Cl_6]$ and synthetic $Ca_{12}Mg_5Si_9O_{32}[Cl_6]$.

The mayenite structure has a high flexibility in element substitution. The crucial substitutions for the classification of hitherto described minerals of the mayenite supergroup are the following:

1. $(Al,Fe^{3+})_T + \square_{[W]} \leftrightarrow Si^{4+}_T + (Cl^-, F^-)_{[W]}$. This substitution relates the mayenite-group with the wadalite-group in a linear series between (Cl, F) = 2, Si = 0 and (Cl,F) = 6 and Si = 4.
2. The substitution $Fe^{3+}_T \leftrightarrow Al_T$, relates wadalite $Ca_{12}Al_{10}Si_4O_{32}[Cl_6]$ and eltyubyuite $Ca_{12}Fe^{3+}_{10}Si_4O_{32}[Cl_6]$ within the wadalite group.
3. The heterovalent substitution $2 \times (Al, Fe^{3+})_T \leftrightarrow (Mg^{2+}+Si^{4+})_T$ relates wadalite/eltyubyuite with the theoretical endmember $Ca_{12}Mg_5Si_9O_{32}[Cl_6]$ within the wadalite group. Adrianite $Ca_{12}Mg_3Al_4Si_7O_{32}[Cl_6]$ [62] has been approved as a new mineral by the nomenclature commission of the International Mineralogical Association (IMA). However, adrianite is actually a solid-solution member between wadalite $Ca_{12}Al_{10}Si_4O_{32}[Cl_6]$ and synthetic $Ca_{12}Mg_5Si_9O_{32}[Cl_6]$ (see 7.7 Synthesis and structure of wadalite-group members).
4. H_2O molecules within the structural cages, $\square_{[W]} \leftrightarrow H_2O_{[W]}$ relate chlorkyuygenite and fluorkyuygenite with chlor- and fluormayenite.

5. Monovalent anions at the [W]-site $F^- \leftrightarrow Cl^-$ define the difference between chlormayenite and fluormayenite or chlorkyuygenite and fluorkyuygenite, respectively.

7.6.2 Achtarandite

Achtarandite is a polycrystalline idiomorphic pseudomorph after a protomineral of the mayenite supergroup [63–65]. The first reported discovery of achtarandite in 1790 goes back to Laksman [66]. Achtarandite pseudomorphs have been found together with large wiluite crystals (vesuvianite group) close to the Achtaranda River mouth, Wiluy River Yakutia, Russia (Fig. 7.7). The pseudomorphs are enclosed in metasomatically altered carbonate-clay rocks (serpentinites), crosscut by diabasic dikes [63]. The formation of the achtarandite bearing rocks is explained with a multiple step metamorphosis. A first high-temperature event, triggered by dikes, is followed by serpentinization. Due to the serpentinization most HT-minerals including the achtarandite protophase have been replaced, resulting in unusual hydrogarnet-serpentinite rocks [63]. Early studies on achtarandite suggested mainly different kinds of garnet as pro-

Fig. 7.7: Large achtarandite pseudomorph (left) with attached wiluite (vesuvianite group) from the Wiluy River deposit, Sakha-Yakutia, Russia.

tophase [63]. However, the tristetrahedral shape of the pseudomorphs suggests point group $\bar{4}3m$ for the protocrystals, which is consistent with minerals of the mayenite supergroup. The actual protomineral is still under discussion, while [64, 65] suggest wadalite, a more recent publication [67] reports chlormayenite as a protophase. Unaltered mayenite supergroup-minerals are commonly rather small and do not exceed millimetre scale. Previously, no comparable fresh achtarandite protomineral has been found.

A careful examination of the mineral association shows that the achtarandite protomineral grew epitaxial on grossular ((112) Ach // (211) or (121) or (112) Grs) [63]. The mineral content of the pseudomorphs is typically dominated by hydrogarnet with minor chlorite, serpentine and occasionally calcite, kaolinite and saponite. The internal structure of achtarandite may be described as a sponge-like skeleton of hydrogarnet, while the other minerals fill the cavities. The crystallographic orientation of the hydrogarnet skeleton is such that the 3-fold axis of the octahedral shaped hydrogarnet crystals coincides with the 3-fold axis of the achtarandite protocrystals. The epitaxy of achtarandite on grossular, and the oriented growth of the hydrogarnet skeleton within achtarandite, is likely caused by the close relation between the mayenite-type and the garnet-type structure.

The conversion of katoite, $Ca_3Al_2(O_4H_4)_3$, to $Ca_{12}Al_{14}O_{32}[O]$, later named mayenite, was the base for the discovery of the title compound. The reverse reaction seems to be the key mechanism for the formation of achtarandite pseudomorphs. The topotactic dehydration and rehydration of $Ca_3Al_2(O_4H_4)_3$ was traced [68] by an X-ray study of single crystals heated in a capillary at 240, 250, and 260 °C. The dehydration proceeded according to the reaction $7(Ca_3Al_2(O_4H_4)_3) \rightarrow Ca_{12}Al_{14}O_{32}[O] + 9Ca(OH)_2 + 33\ H_2O$. Continuous conversion has been observed from the crystal surface to the core, since sharp X-ray reflections of educts and products coexisted for several hours. During the rehydration of the once dehydrated sample an additional intermediate phase (IP) $Ca_{12}[^{VI}Al(OH)_6]_8[^{IV}AlO_3(OH)]_6$ formed. Structurally the IP may be derived from the mayenite structure, whereby all O2 sites are replaced by $3 \times OH^-$ similar to the hydroxylation of chlor- and fluormayenite. To preserve charge balance six of 24 O1 sites pfu are replaced by OH^-. The reaction is described in two steps as:

1. $Ca_{12}Al_{14}O_{32}[O]$ (mayenite) + 15 $H_2O \rightarrow Ca_{12}[Al(OH)_6]_8[AlO_3(OH)]_6$ (IP)
2. $Ca_{12}[Al(OH)_6]_8[AlO_3(OH)]_6$ (IP) + 18 $H_2O \rightarrow 4\ Ca_3Al_2(O_4H_4)_3$ (katoite) + 4 $Al(OH)_3$.

The uptake of 15 H_2O is needed to retain the garnet structure which was observed for the (IP) [68]. Knowing from natural samples that the partial hydroxylation of the O2 site in the mayenite-type structure may occur in a disordered fashion, the question arises whether a gradual hydroxylation series between mayenite and (IP) exists. A two-step substitution mechanism can be postulated:

1. $O^{2-}_{(O2)} + ½\, O^{2-}_{([W])} \rightarrow 3 \times OH^-_{(O2a)}$ relating $Ca_{12}Al_{14}O_{32}[O]$ to $Ca_{12}Al_{14}O_{30}(OH)_6$
2. $O^{2-}_{(O2)} + O^{2-}_{(O1)} \rightarrow 3 \times OH^-_{(O2a)} + OH^-_{(O1)}$ relating $Ca_{12}Al_{14}O_{30}(OH)_6$ to (IP)

Surprisingly, the oriented hydroxylation has only been shown for previously dehydroxylated katoite $Ca_3Al_2(O_4H_4)_3$. However, the preservation of the crystallographic orientation suggests a member of the mayenite supergroup as protomineral of achtarandite. The presence of abundant silicates (chlorite, serpentine, Si-bearing hydrogarnet) within the achtarandite pseudomorphs favours a wadalite-like mineral as a protophase. However, this issue remains ambiguous.

7.6.3 Silicates with mayenite-type structure

Although Si-bearing mayenite-type compounds like $Ca_{12}Al_{10}Si_4O_{32}[O_3]$ derived from hydrogarnet are currently discussed in material science [50, 51], the interest of the scientific community in silicates with mayenite structure is less enthusiastic than for mayenite *sensu stricto*. Melt grown silicates with mayenite-type structure do not show promise for mayenite-like technological applications as all structural cages are commonly occupied by monovalent anions such as Cl. The first reported compound was $Ca_{12}Al_{10.6}Si_{3.4}O_{32}Cl_{5.4}$, grown from a mixture of CaO, SiO_2, Al_2O_3 and excess $CaCl_2$ in open crucibles at 1123 K for 72 h [69]. Other reports on synthetic wadalite-like phases are application related. For instance, silicates with mayenite-type structure are discussed as solids to immobilize toxic elements from Cl-rich media, namely the radioactive fission products from spent nuclear fuel after reprocessing in chloride melt [70, 71]. Another example is formation of a wadalite-like phase during HCl removal from effluent gases, by reaction with hydrogrossular [72, 73].

The natural analogues belong to the wadalite-group and are characteristic for Cl-bearing mineral associations in altered carbonate xenoliths enclosed in volcanic rocks. Known sources are a quarry in Tadona near Koriyama City, Fukushima Prefecture, Japan [74], the La Negra mine, Queretaro, Mexico [74], the Bellerberg Volcano, Eifel, Germany [76], the Upper Chegem Caldera, Northern Caucasus [77], and the Shadil-Khokh volcano, Southern Ossetia [78]. In addition to the terrestrial occurrences, wadalite is also found in calcium-aluminium-rich inclusions of meteorites (e.g. [79]).

When introducing the mayenite supergroup [61] the existence of Mg-rich wadalite was discussed for the first time. Synthesis and crystal structures of a series of Mg-rich compositions between $Ca_{12}Al_{10}Si_4O_{32}[Cl_6]$, $Ca_{12}Fe_{10}Si_4O_{32}[Cl_6]$, and $Ca_{12}Mg_5Si_9O_{32}[Cl_6]$ is presented in the following paragraphs.

7.7 Synthesis and structure of wadalite-group members

The synthesis of Mg-rich wadalites has been performed using a $CaCl_2$-flux. Mixtures of pure oxides SiO_2 (Sigma-Aldrich S5631), Al_2O_3 (Sigma-Aldrich 23474-5), MgO (Sigma-Aldrich 243388), Fe_2O_3 (Sigma-Aldrich 310050) and Ca-carbonate (Sigma-Aldrich

239216) were degassed and presintered at 1000 °C for 6h in Pt-crucibles. $CaCl_2 \cdot 2H_2O$ (Sigma-Aldrich 223506) was dehydrated at 200 °C and subsequently added in excess to the pre-processed mixtures. Two types of reactors were tested: (1) Pt-crucibles with lids and (2) welded Pt-tubes. In both cases the reactors were heated to 1200 °C within 3h and subsequently cooled to 780 °C with a cooling rate of 10 °C/h. Alternatively a cooling rate of 5 °C/h has been tested with similar results. Both reactors are suitable for the synthesis, whereby crucibles suffer evaporation of Cl. The expected reactions for the here presented compounds were:

1. $9CaO + 5MgO + 9SiO_2 + 3CaCl_2 \rightarrow Ca_{12}(Mg_5Si)(Si_8)O_{32}Cl_6$
2. $9CaO + Fe_2O_3 + 4MgO + 8SiO_2 + 3CaCl_2 \rightarrow Ca_{12}(Mg_4Fe_2)(Si_8)O_{32}Cl_6$
3. $9CaO + Al_2O_3 + 4MgO + 8SiO_2 + 3CaCl_2 \rightarrow Ca_{12}(Mg_4Al_2)(Si_8)O_{32}Cl_6$
4. $9CaO+0.5\,Al_2O_3+0.5Fe_2O_3+4MgO + 8SiO_2+3CaCl_2 \rightarrow Ca_{12}(Mg_4FeAl)(Si_8)O_{32}Cl_6$

Products are abbreviated "WA" (wadalite) and the cation species occupying the T-sites are denoted by subscripts, e.g. $WA_{(MgSi)}$, $WA_{(MgFeSi)}$, $WA_{(MgAlSi)}$ and $WA_{(MgAlFeSi)}$.The products are idiomorphic single crystals from µm- up to mm-scale, examples are shown in Fig. 7.8. Besides silicates with mayenite-type structure, all runs showed additional by-products, mainly compounds of the melilite group, $Ca_2(Al,Mg,Fe)(Si,Al)_2O_7$, monticellite, $CaMg(SiO_4)$, diopside, $CaMgSi_2O_6$, and anorthite, $CaAl_2Si_2O_8$. Moreover, unreacted and recrystallized Fe_2O_3 was observed in reaction (2) and (4). $WA_{(MgSi)}$ and $WA_{(MgAlSi)}$ are colourless transparent while Fe-bearing samples are yellow-translucent. In addition, Fe-bearing samples may appear non-transparent orange due micro-inclusions of hematite Fe_2O_3. Multiple samples of all kinds have been measured on a Bruker APEXII single-crystal diffractometer. The CIF files including details about the X-ray measurements have been submitted to ICSD FIZ Karlsruhe.

Since the scattering factors of Mg, Al and Si are too similar for quantitative distinction, the occupancy factors of the T-sites for $WA_{(MgSi)}$ and $WA_{(MgAlSi)}$ are based on bond-lengths and charge balance. In $WA_{(MgFeSi)}$, and $WA_{(MgAlFeSi)}$ site occupancies for $T1$ have been refined with the constraint occupancy (Si_{T1}) $= x$ and occupancy (Fe_{T1}) $= 1 - x$. The site occupancy of $T2$ Fe was refined against Mg as a first step. In a second step the refined value for Fe was fixed and the residual electrons were assigned to Mg/Si or Mg/Al in order to maintain charge balance. In $WA_{(MgSi)}$, $WA_{(MgFeSi)}$, and $WA_{(MgAlFeSi)}$ the occupancy factor for Cl at $[W]$ has been refined.

As discussed above, the $T1$-site in mayenite-type compounds corresponds to the octahedral site, and $T2$ to the tetrahedral Si-site in garnets. At first glance this analogy to garnets would suggest incorporating the bigger cations Mg, Al, Fe^{3+} preferentially at the $T1$ site. O2, the fourth apex of the $T1$ tetrahedron is however only bonded to one tetrahedral cation and 3 × Ca1. According to Pauling's valence rule Mg^{2+} at a tetrahedral site contributes only 0.5 valences to each coordinating O site, thus O2 would likely be underbonded. This assumption is confirmed by all measured Mg-wadalites. The average bond length for $T1$ is 1.636 Å in $WA_{(MgSi)}$ and 1.640 Å in $WA_{(MgAlSi)}$. In both cases $T1$ has been considered as a pure SiO_4

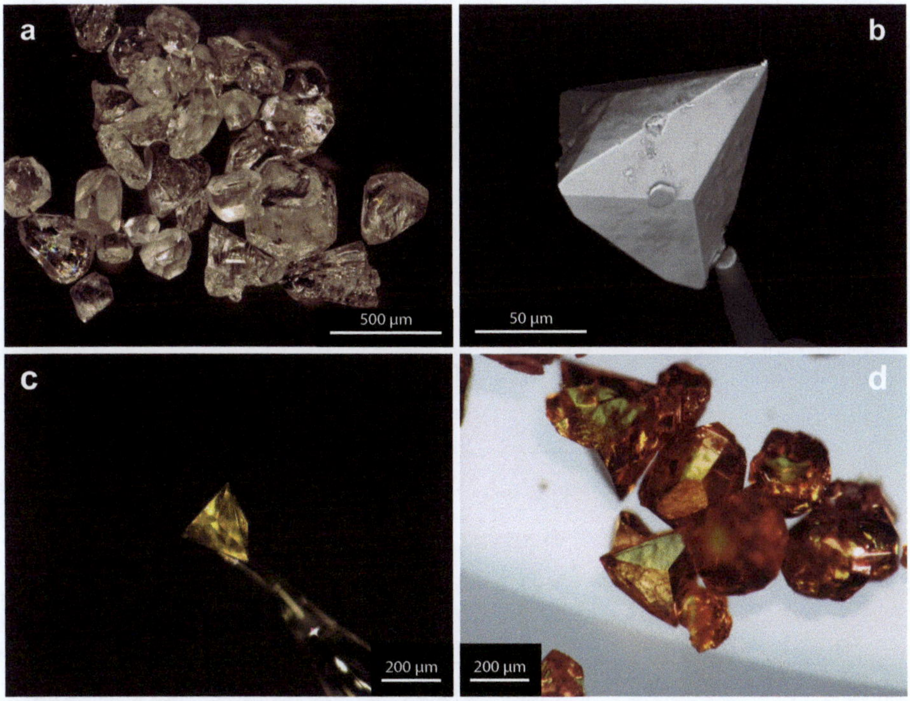

Fig. 7.8: Four examples of synthetic Mg-wadalites, (a) Light-microscopy picture of $Ca_{12}(Mg_4Al_2)(Si_8)O_{32}Cl_6$. (b) Backscattered-electron picture of idiomorphic $Ca_{12}(Mg_5Si)(Si_8)O_{32}Cl_6$. (c) idiomorphic $Ca_{12}(Mg_4FeAl)(Si_8)O_{32}Cl_6$ crystal glued on a glass needle. (d) single crystals of $Ca_{12}(Mg_4Fe_2)(Si_8)O_{32}Cl_6$.

tetrahedron though minor amounts of Al in $WA_{(MgAlSi)}$ are possible. In $WA_{(MgFeSi)}$ and $WA_{(MgAlFeSi)}$ *T*1 is also Si dominant with slightly longer average bond lengths of 1.645 Å and 1.646 Å. The refined site occupancy factors for Fe at *T*1 converged to 0.049(2) ($WA_{(MgFeSi)}$) and 0.043(3) ($WA_{(MgAlFeSi)}$). The *T*2 site is in all cases Mg-dominant. In $WA_{(MgSi)}$ *T*2 is shared by Mg and Si leading to a refined formula of $Ca_{12}(Mg_{5.125}Si_{0.875})(Si_8)O_{32}Cl_{5.74(18)}$ with a *T*2-O1 bond length of 1.8532(8) Å. As [*W*] is fully occupied by Cl in $WA_{(MgAlSi)}$ the *T*2 site is shared by 2/3 Mg and 1/3 Al, leading to the formula $Ca_{12}(Mg_4Al_2)(Si_8)O_{32}Cl_6$ with *T*2-O1 bond length of 1.8371(8) Å. In $WA_{(MgFeSi)}$ of $Ca_{12}(Mg_{4.08}Fe_{1.6}Si_{0.32})(Si_{7.61(16)}Fe_{0.39(16)})O_{32}Cl_{5.86(18)}$ composition *T*2 is shared by Mg, Fe and minor Si with *T*2-O bond lengths of 1.8707(8) Å. The lowest Mg content at *T*2 is observed for crystals of the $WA_{(MgAlFeSi)}$-series. As aimed by the starting compositions, *T*2 is shared by the three cations Al, Fe, Mg. The sample presented here shows a refined formula of $Ca_{12}(Mg_{3.71}Fe_{0.7}Al_{1.59})(Si_{7.66(2)}Fe_{0.34(2)})O_{32}[Cl_{5.88(18)}]$ and a *T*2-O1 bond length of 1.8496(8) Å. It has to be stated that the $WA_{(MgAlFeSi)}$- and $WA_{(MgFeSi)}$-series show considerable chemical variation as shown in Fig. 7.9.

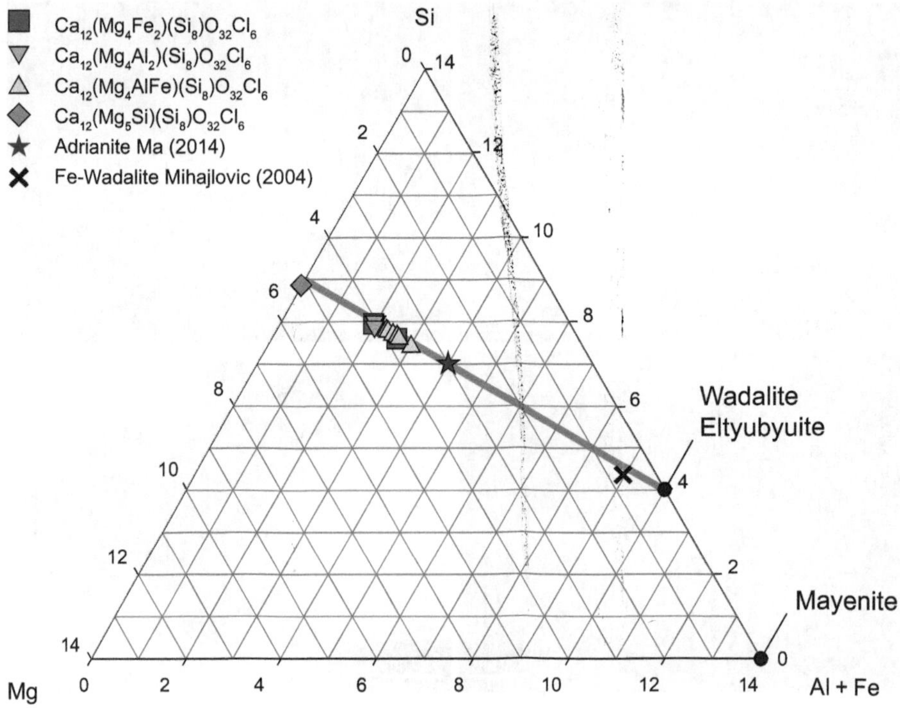

Fig. 7.9: Ternary diagram normalized to 14 tetrahedral sites of the mayenite type structure. The grey bar indicates the substitution mechanism $2 \times (Al,Fe^{3+})_T \leftrightarrow (Mg^{2+} + Si^{4+})_T$ connecting synthetic Mg-wadalite with wadalite and eltyubyuite. The series $WA_{(MgFeSi)}$ (red) and $WA_{(MgAlFeSi)}$ (yellow) show a considerable scatter along the line. Most compounds are slightly below the theoretical line due to partial occupancy of [*W*].

It is noteworthy that in all cases except $WA_{(MgAlSi)}$ the [*W*]-site remains only partially occupied by Cl. This observation is in agreement with all hitherto published single-crystal data of wadalite-group minerals and analogous compounds. The presented refinements indicate a complete solid solution series between MgSi-wadalite, eltyubyuite $Ca_{12}(Fe_6)(Si_4Fe_4)O_{32}[Cl_6]$ and wadalite. While Mg^{2+} is restricted to the *T*2 site, Al and Fe^{3+} may be incorporated at either tetrahedral site. Si shows a slight preference for *T*1 but may, however, occupy both *T*-sites. Regarding the difference in ionic radii of $^{IV}Si^{4+}$ (0.26) and $^{IV}Mg^{2+}$ (0.57) [80] a mixed Mg/Si occupancy at a *T*-site is surprising.

7.8 Outlook

A microporous crystal structure allowing for anion diffusion and chemical diversity makes mayenite an excellent object for present and future research with a wide range of potential technical applications. An insufficiently explored issue is the arrangement and long-range order of cage-filling oxygen O3, which is only partially occupied and responsible for anion conductivity. Since O3 interacts with the framework by breaking an Al1-O1 bond, the absolute site for the adjacent O3 may not be at random. Conventional diffraction data deliver only time- and space-averaged information with the constraint of overall cubic symmetry. However, in reality we would expect some ordering or domain formation, which may experimentally be tackled by time-resolved spectroscopic techniques and theoretically by structure modelling tools.

An interesting project in mineralogical crystallography could be a precise in-situ X-ray study on the hydroxylation of mayenite with focus on the conversion from the mayenite to the hydrogarnet structure. Moreover, there are likely still undescribed new minerals of the mayenite supergroup to be found and analysed. Their mode of occurrence may provide new synthesis routes and their composition and properties may inspire scientists to find additional applications. The flux-based synthesis of wadalite-like phases opens a wide range of possible compounds providing the opportunity for a deeper understanding of the crystal chemistry of silicates with mayenite-type structure.

Acknowledgment: The author thanks the reviewers, who provided helpful comments on earlier drafts of the manuscript and helped to ameliorate the English writing. This study was supported by the Swiss National Science Foundation (project "crystal chemistry of minerals" Nr. 200020_134617).

Bibliography

[1] Shepherd ES, Rankin GS. The binary systems of alumina with silica, lime, and magnesia. Am J Sci 1909, 166, 293–333.

[2] Köberich F. Die Maximaltension und das Adsorptionsgleichgewicht: ein Beitrag zur Kinetik heterogener Gleichgewichte. Diss. Univ. Berlin, Mitt. Kaiser Wilhelm Inst. Silikatforschung 1934, 154

[3] Lager GA, Armbruster T, Faber J. Neutron and X-ray diffraction study of hydrogarnet $Ca_3Al_2(O_4H_4)_3$. Am Mineral 1987, 72, 756–765.

[4] Büssem W. Eitel A. Die Struktur des Pentacalciumtrialuminats. Z Kristallogr 1936, 95,175–188.

[5] Jeevaratnam J, Glasser, LD, Glasser FP. Structure of calcium aluminate, 12CaO. $7Al_2O_3$. Nature, 1962, 194, 764–765

[6] Bartl, H. Roentgen-Einkristalluntersuchungen an $3(CaO)\bullet Al_2O_3\bullet(H_2O)_6$ und 12 (CaO)•7 $(Al_2O_3)\bullet(H_2O)$. Neuer Vorschlag Zur $12(CaO)\bullet 7(Al_2O_3)$-Struktur. N Jb Miner Mh 1969, 404–412.

[7] Kuzel, HJ. (1969). Über die orientierte Entwässerung von Tricalciumaluminathydrat $3CaO\bullet Al_2O_3\bullet 6H_2O$. N Jb Miner Mh 1969, 397–404.

[8] Jeevaratnam J, Glasser FP, Dent Glasser LS. Anion substitution and structure of $12CaO\bullet 7Al_2O_3$. J Am Ceram Soc 1964, 47, 105–106.

[9] Williams PP. Crystal structure of a fluoride derivative of $12CaO\bullet7Al_2O_3$. J Am Ceram Soc 1968, 51, 531–531.
[10] Williams PP. Refinement of the structure of $11CaO\bullet7Al_2O_3\bullet CaF_2$. Acta Crystallogr B 1973, 29(7), 1550–1551.
[11] Sushko PV, Munoz Ramo D, Shluger AL. Electronic structure and spectroscopic properties of interstitial anions in the nanoporous complex oxide $12CaO\bullet7Al_2O_3$. Phys Status Solidi A. 2007, 204, 663–669.
[12] Boysen H, Lerch M, Stys A, Senyshyn A. Structure and oxygen mobility in mayenite ($Ca_{12}Al_{14}O_{33}$): a high-temperature neutron powder diffraction study. Acta Crystallogr B 2007, 63, 675–682.
[13] Palacios L, De La Torre AG, Bruque S, García-Muñoz JL, García-Granda S, Sheptyakov D, Aranda MA. Crystal structures and in-situ formation study of mayenite electrides. Inorg Chem 2007, 46, 4167–4176.
[14] Palacios L, Cabeza A, Bruque S, García-Granda S, Aranda MA. Structure and electrons in mayenite electrides. Inorg Chem 2008, 47, 2661–2667.
[15] Sakakura T, Tanaka K, Takenaka Y, Matsuishi S, Hosono H, Kishimoto S. Determination of the local structure of a cage with an oxygen ion in $Ca_{12}Al_{14}O_{33}$. Acta Crystallogr B 2011, 67(3), 193–204.
[16] Hayashi K, Hirano M, Hosono H. Thermodynamics and kinetics of hydroxide ion formation in $12CaO\bullet7Al_2O_3$. J Phys Chem B 2005, 109, 11900–11906.
[17] Strandbakke R, Kongshaug C, Haugsrud R, Norby T. High-temperature hydration and conductivity of mayenite, $Ca_{12}Al_{14}O_{33}$. J Phys Chem C 2009, 113, 8938–8944.
[18] Tolkacheva AS, Shkerin SN, Korzun IV, Plaksin SV, Khrustov VR, Ordinartsev DP. Phase transition in mayenite $Ca_{12}Al_{14}O_{33}$. Russ J Inorg Chem 2012, 57, 1014–1018.
[19] Brunner GO, Meier WM. Framework density distribution of zeolite-type tetrahedral nets. Nature 1989, 337, 146–147.
[20] Loewenstein, W. The distribution of aluminum in the tetrahedra of silicates and aluminates. Am Mineral 1954, 39, 92–96.
[21] Schmidt A, Boysen H, Senyshyn A, Lerch M. New findings on N-mayenite and a new kind of anion substituted mayenite: $Ca_{12}Al_{14}O_{32}(NO_2)_2$. Z Kristallogr 2014, 229(6), 427–434.
[22] Ingram BJ, Gonzalez GB, Kammler DR, Bertoni MI, Mason TO. Chemical and structural factors governing transparent conductivity in oxides. J Electroceram 2004, 13(1–3), 167–175.
[23] Hosono H, Hayashi K, Hirano M. Active anion manipulation for emergence of active functions in the nanoporous crystal $12CaO\bullet7Al_2O_3$: a case study of abundant element strategy. J Mater Sci 2007, 42(6), 1872–1883.
[24] Yang S, Kondo JN, Hayashi K, Hirano M, Domen K, Hosono H. Partial oxidation of methane to syngas over promoted $C_{12}A_7$. Appl Catal A 2004, 277(1), 239–246.
[25] Li C, Hirabayashi D, Suzuki K. A crucial role of O_2^- and O_2^{2-} on mayenite structure for biomass tar steam reforming over $Ni/Ca_{12}Al_{14}O_{33}$. Appl Catal B 2009, 88(3), 351–360.
[26] Dye JL. Electrons as anions. Science 2003, 301(5633), 607–608.
[27] Markham JJ. F-centers in Alkali Halides (Vol. 8). Academic Press Inc. NY, USA, 1966.
[28] Huang RH, Wagner MJ, Gilbert DJ, Reidy-Cedergren KA, Ward DL, Faber MK, Dye JL. Structure and properties of Li^+(Cryptand [2.1.1])e^-, an electride with a 1D “spin-ladder-like” cavity-channel geometry. J Am Chem Soc 1997, 119, 3765–3772.
[29] Ward DL, Huang RH, Dye JL. Structures of alkalides and alectrides. I. structure of potassium cryptand[2.2.2]* electride. Acta Crystallogr C 1988, 44, 1374–1376.
[30] Dawes SB, Ward DL, Huang RH, Dye JL. First electride crystal structure. J Am Chem Soc 1986, 108, 3535–3537.

[31] Edwards PP, Anderson PA, Woodallx LJ, Porch A, Armstrong AR. Can we synthesise a dense bundle of quasi one-dimensional metallic wires? Mat Sci Eng A 1996, 217, 198–202.

[32] Srdanov VI, Haug K, Metiu H, Stucky GD. Sodium (Na_4^{3+}) clusters in sodium sodalite. J Phys Chem 1992, 96(22), 9039–9043.

[33] Hayashi K, Matsuishi S, Kamiya T, Hirano M, Hosono H. Light-induced conversion of an insulating refractory oxide into a persistent electronic conductor. Nature 2002, 419,462–465.

[34] Matsuishi S, Toda Y, Miyakawa M, Hayashi K, Hirano M, Tanaka I, Hosono H. High-density electron anions in a nanoporous single crystal:$[Ca_{24}Al_{28}O_{64}]^{4+}$ ($4e^-$). Science 2003, 301, 626.

[35] Kim S, Miyakawa M, Hayashi K, Sakai T, Hirano M, Hosono H. Simple and efficient fabrication of room temperature stable electride: Melt-solidification and glass ceramics. J Am Chem Soc 2005, 127, 1370–1371.

[36] Kim SW, Matsuishi S, Miyakawa M, Hayashi K, Hirano M, Hosono H. Fabrication of room temperature-stable $12CaO•7Al_2O_3$ electride: a review. J Mater Sci 2007, 18, 5–14.

[37] Pereira EG, da Silva JN, de Oliveira JL, Machado CS. Sustainable energy: a review of gasification technologies. Renew Sust Energ Rev 2012, 16, 4753–4762.

[38] Yung MM, Jablonski WS, Magrini-Bair KA. Review of catalytic conditioning of biomass-derived syngas. Energ Fuel 2009, 23, 1874–1887.

[39] Che M, Tench AJ. Characterization and reactivity of molecular oxygen species on oxide surfaces. Adv Catal 1983, 32, 1–148.

[40] Foote CS. (Ed.). Active oxygen in chemistry. Springer Science & Business Media 1995.

[41] Valentine JS, Foote CS, Greenberg A, Liebman, JF. Active oxygen in biochemistry. New York, NY: Blackie Academic & Professional 1995.

[42] Hayashi K, Hirano M, Matsuishi S, Hosono H. Microporous Crystal $12CaO•7Al_2O_3$ Encaging Abundant O-Radicals. J Am Chem Soc 2002, 124, 738–739.

[43] Hayashi K, Matsuishi S, Ueda N, Hirano M, Hosono H. Maximum Incorporation of Oxygen Radicals, O^- and O_2^-, into $12CaO\ 7Al_2O_3$ with a nanoporous structure. Chem Mater 2003, 15, 1851–1854.

[44] Lemonidou AA, Vasalos IA. Preparation and evaluation of catalysts for the production of ethylene via steam cracking: effect of operating conditions on the performance of 12CaO-$7Al_2O_3$ Catalyst. Appl Catal 1989, 54, 119–138.

[45] Pant KK, Kunzru D. Catalytic pyrolysis of n-heptane: kinetics and modeling. Ind Eng Chem 1997, 36, 2059–2065.

[46] Pant KK, Kunzru D. Catalytic pyrolysis of methylcyclohexane: kinetics and modeling. Chem Eng J 1998, 70, 47–54.

[47] Ruszak M, Inger M, Witkowski S, Wilk M, Kotarba A, Sojka Z. Selective N_2O removal from the process gas of nitric acid plants over ceramic $12CaO•7Al_2O_3$ catalyst. Catal Lett 2008, 126, 72–77.

[48] Li QX, Hayashi K, Nishioka M, Kashiwagi H, Hirano M, Torimoto Y, Hosono H, Sadakata M. Absolute emission current density of O-from $12CaO•7Al_2O_3$ crystal. Appl Phys Lett 2002, 80, 4259–4261.

[49] Schmidt DCA. Synthese und Charakterisierung substituierter Mayenitphasen. Dissertation Technische Universität Berlin 2014

[50] Fujita S, Suzuki K, Ohkawa M, Mori T, Iida Y, Miwa Y, Masuda H, Shimada S. Oxidative destruction of hydrocarbons on a new zeolite-like crystal of $Ca_{12}Al_{10}Si_4O_{35}$ including O_2^- and O_2^{2-} radicals. Chem Mater 2003, 15, 255–263.

[51] Bertoni MI, Mason TO, Medvedeva JE, Wang Y, Freeman AJ, Poeppelmeier KR. Enhanced electronic conductivity in Si-substituted calcium aluminate. J Appl Phys 2007, 102, 113704–113704.

[52] Jasmund K, Hentschel G. Seltene Mineralparagenesen in den Kalksteineinschlüssen der Lava des Ettringer Bellerberges bei Mayen (Eifel). Beitr Mineral Petrol 1964, 10, 296–314.
[53] Lengauer CL, Tillmanns E, Ntaflos T, Hentschel G. Neue Mineralfunde vom Bellerberg, Eifel. Eur J Mineral 1997, 9, Bh1, 220.
[54] Hentschel G. Mayenite, $12CaO•7Al_2O_3$, and brownmillerite, $2CaO•(Al, Fe)_2O_3$, two new minerals in limestone inclusions in the lava of the Ettringer Bellerberg. N Jb Miner Mh 1964, 22–29.
[55] Hentschel, G. Die Mineralien der Eifelvulkane. Weise Verlag, München (2. Auflage), 1987.
[56] Gross S. The mineralogy of the Hatrurim Formation, Israel. Bull Geol Sur Isr 1977, 70, 1–80.
[57] Heritsch H. Eine Kontaktbildung aus dem Nephelinbasanit Steinbruch von Klöch (Südoststeiermark) mit seltenen Mineralien; natürliches Vorkommen der Verbindung $4CaO•3Al_2O_3•SO_3$. Mitt Abt Mineral Land Joan 1990, 58, 15–35.
[58] Ma C, Connolly HC, Beckett JR, Tschauner O, Rossman GR, Kampf AR, Zega TJ, Smith SAS, Schrader DL. Brearleyite, $Ca_{12}Al_{14}O_{32}Cl_2$, a new alteration mineral from the NWA 1934 meteorite. Am Mineral 2011, 96, 1199–1206.
[59] Galuskin EV, Kusz J, Armbruster T, Bailau R, Galuskina IO, Ternes B, Murashko M. A reinvestigation of mayenite from the type locality, the Ettringer Bellerberg volcano near Mayen, Eifel district, Germany. Mineral Mag 2012, 76, 707–716.
[60] Galuskin EV, Gfeller F, Armbruster T, Galuskina IO, Vapnik Y, Dulski M, Murashko M, Dzierżanowski P, Sharygin VV, Krivovichev SV, Wirth, R. Mayenite supergroup, Part III: Fluormayenite, $Ca_{12}Al_{14}O_{32}$ [$\square_4F_2$], and fluorkyuygenite, $Ca_{12}Al_{14}O_{32}$ [$(H_2O)_4F_2$], two new minerals from pyrometamorphic rocks of the Hatrurim Complex, South Levant. Eur J Mineral 2015, 27, 123–136,
[61] Galuskin EV, Gfeller F, Galuskina IO, Armbruster T, Bailau R, Sharygin, V. V. Mayenite supergroup, part I. Recommended nomenclature. Eur J Mineral 2015, 27, 99–111
[62] Ma C, Krot AN. Adrianite, IMA 2014-028. CNMNC Newsletter No. 21, August 2014, page 801, Mineral Mag 2014, 78, 797–804.
[63] Galuskin E, Galuskina I, Winiarska A. Epitaxy of achtarandite on grossular - the key to the problem of achtarandite. N Jb Miner Mh 1995, 306–320.
[64] Galuskina I, Galuskin E, Sitarz M. Atoll hydrogarnets and mechanism of the formation of achtarandite pseudomorphs. N Jb Miner Mh 1998, 49–62.
[65] Galuskin E, Galuskina I. Achtarandite–sponge hibschite pseudomorph after wadalite-like phase: internal morphology and mechanism of formation. N Jb Miner Abh 2002, 178, 63–74.
[66] Lyakhovich VV. New data on mineralogy of the Vilyui achtarandite deposit. Works of the East-Siberian Branch of the Academy of Science USSR 1954, 1.
[67] Alferova MS. Grossular, vesuvianite and achtarandite mineralogy of the Talnakh Region. New Data Mineral 2007, 42, 50–61
[68] Schröpfer L, Bartl H. Oriented decomposition and reconstruction of hydrogarnet, $Ca_3Al_2(OH)_{12}$. Eur J Mineral 1993, 5, 1133–1144.
[69] Feng QL, Glasser FP, Howie RA, Lachowski EE. Chlorosilicate with the $12CaO•7Al_2O_3$ structure and its relationship to garnet. Acta Crystallogr C 1988, 44, 589–592.
[70] Leturcq G, Grandjean A, Rigaud D, Perouty P, Charlot M. Immobilization of fission products arising from pyrometallurgical reprocessing in chloride media. J. Nucl Mater 2005, 347, 1–11.
[71] Park HS, Kim IT, Cho YZ, Eun HC, Lee HS. Stabilization/solidification of radioactive salt waste by using x SiO_2 – y Al_2O_3 – z P_2O_5 (SAP) material at molten salt state. Environm Sci Technol 2008, 42, 9357–9362.
[72] Fujita S, Suzuki K, Ohkawa M, Shibasaki Y, Mori T. Reaction of hydrogrossular with hydrogen chloride gas at high temperature. Chem Mater 2001, 13, 2523–2527.
[73] Fujita S, Suzuki K, Mori T, Shibasaki Y. A new technique to remove hydrogen chloride gas at high temperature using hydrogrossular. Ind Eng Chem 2003, 42, 1023–1027.

[74] Tsukimura K, Kanazawa Y, Aoki M, Bunno M. Structure of wadalite $Ca_6Al_5Si_2O_{16}Cl_3$, Acta Crystallogr 1993, C49, 205–207.

[75] Kanazawa Y, Aoki M, Takeda H. Wadalite, rustumite, and spurrite from La Negra mine, Queretaro, Mexico. Bull Geol Sur Jpn 1997, 48, 413–420.

[76] Mihajlović T, Lengauer CL, Ntaflos T, Kolitsch U, Tillmanns E. Two new minerals, rondorfite, $Ca_8Mg[SiO_4]_4Cl_2$, and almarudite, $K(\square, Na)_2(Mn,Fe,Mg)_2(Be,Al)_3[Si_{12}O_{30}]$, and a study of iron-rich wadalite, $Ca_{12}[(Al_8Si_4Fe_2)O_{32}]Cl_6$, from the Bellerberg (Bellberg) volcano, Eifel. N Jb Miner Abh 2004, 179, 265–294.

[77] Galuskin EV, Galuskina IO, Bailau R, Prusik K, Gazeev VM, Zadov AE, Pertsev NN, JeżakL, Gurbanov AG, Dubrovinsky L. Eltyubyuite, $Ca_{12}Fe^{3+}_{10}Si_4O_{32}Cl_6$ – the Fe^{3+} analogue of wadalite: A new mineral from the Northern Caucasus, Kabardino-Balkaria, Russia. Eur J Mineral 2013, 25, 221–229.

[78] Gfeller F, Środek D, Kusz J, Dulski M, Gazeev V, Galuskina IO, Galuskin EV, Armbruster T. Mayenite supergroup, part IV: Crystal structure and Raman investigation of Al-free eltyubyuite from the Shadil-Khokh volcano, Kel'Plateau, Southern Ossetia. Eur J Mineral 2015, 27, 137–143

[79] Ishii HA, Krot AN, Bradley JP, Keil K, Nagashima K, Teslich N, Jacobsen B, Yin QZ. Discovery, mineral paragenesis, and origin of wadalite in a meteorite. Am Mineral 2010, 95, 440–448.

[80] Shannon RT. Revised effective ionic radii and systematic studies of interatomic distances in halides and chalcogenides. Acta Crystallogr A 1976, 32, 751–767.

Index

ab initio calculations 159, 170
– ab initio molecular dynamics 87
– Born–Oppenheimer 87
– Car–Parrinello 87
achtarandite 185–187
adrianite 184
advanced ceramics 179–181
ahrensite 129–133, 136–140
$Al_{63}Cu_{24}Fe_{13}$ 112
$Al_{71}Ni_{24}Fe_5$ 120
algorithmic complexity 39–41, 47–48
alunite 50
American Mineralogist Crystal Structure Database (AMCSD) 3, 8–9, 12, 14
anthropogenic processes 116
antigorite 50, 56
anti-zeolite 178
aperiodic structure 111
approximant structure 123
ascent velocity 141–142
automorphism group 38

bakhchisaraitsevite 66
band theory 79
basal plane 99
basis set 86
beam damage 153
bilinite 59–60
biologically-controlled minerals (BCM) 149, 157, 159–161
biologically-induced minerals (BIM) 149, 157, 161
biomineral characterization 149–151, 155–162
biominerals 149–162
bond lengths 91, 133, 172, 188–189
bouazzerite 50, 66
bridgmanite 129, 135, 141, 143
Brillouin 135–136, 138
bulk modulus 135–138, 140–141

$Ca_{12}Al_{10}Si_4O_{32}[Cl_6]$ 184
$Ca_{12}Al_{14}O_{32}[\square_4Cl_2]$ 184
$Ca_{12}Al_{14}O_{32}[\square_4F_2]$ 184
$Ca_{12}Al_{14}O_{32}[\square_5O]$ 169–178
$Ca_{12}Al_{14}O_{32}[(H_2O)_4Cl_2]$ 184
$Ca_{12}Al_{14}O_{32}[(H_2O)_4F_2]$ 184
$Ca_{12}Fe^{3+}_{10}Si_4O_{32}[Cl_6]$ 184
$Ca_{12}Mg_3Al_4Si_7O_{32}[Cl_6]$ 184
$Ca_{12}Mg_5Si_9O_{32}[Cl_6]$ 184, 187–190
$CaCl_2$-flux 187–188
cacoxenite 66
calaverite 110
calcium aluminum inclusions 118
calcium carbonate 157–159
calcium carbonate polymorphism 157
carbonaceous chondrite 118–119
$CaSiO_3$ 128–129
catalysis 180–181
cell parameters 5, 8–9, 23–24
cellular automata 40
charoite 52–53
chemical analysis 6
CheMin 24–25
chem-sorption 93–94
chlorkyuygenite 182, 184–185
chlormayenite 182–184
chrysotile 56
Chukotka 113, 119
clay minerals 57, 94, 99, 155
clusters 110, 112
collision 116
color 1, 15
composite structures 110
computational 135–136, 140, 148
conductivity 179–180, 191
copiapite 59–60
coquimbite 50, 59–60
core boundary 116, 123
cristobalite 43–48, 54, 56, 70
crystal structure 23, 129–130, 132–133, 137, 141, 145
CrystalSleuth 8, 10–11, 21–23
cubic 33, 38, 40–41, 96, 128–131, 169, 191
cupalite 113
cymrite 50, 56, 71

decagonal symmetry 121–123
density functional theory 83–87
– homogeneous electron gas 83–84

– Kohn–Sham equation 84
–DFT+U 85, 93
deterministic finite automaton 43–44
diamond 91–92, 140–144
diffraction 1–15, 21, 24, 75, 109, 113, 115, 123, 133, 135–136, 138, 140, 149, 157, 169, 171
diffraction pattern 5, 8–9, 111–117, 122, 151–154
diopside 50, 113, 128–129, 188
direct electron detectors 162
direct methods 154
dmisteinbergite 48, 55–56
doping 181
dyamical scattering 151
edge sites 99

electrical double layer 90
electride 179–180
electron density 77–79, 82–86, 171–172, 176
electron diffraction 111, 115, 122, 149–162
– automated diffraction tomography (ADT) 151–162
– convergent beam electron-diffraction (CBED) 151, 160, 162
– tomography 151–162
– energy-filtered 162
– Kikuchi lines 151
– nano-electron diffraction (NED) 152
– precession electron diffraction (PED) 153–154
– rotation electron-diffraction (RED) 162
– selected area electron diffraction (SAED) 151, 159
electron emitters 180
electron energy loss spectroscopy (EELS) 150
electron holography (EH) 150
electron-microprobe analysis (EPMA) 6
electronic states 79, 83, 93
eltyubyuite 182, 184, 190
energy dispersive X-ray spectroscopy (EDX, EDS) 150, 155
energy magnetic chiral dichroism (EMCD) 150
equation of state 135
eudialyte 50
exchange and correlation functional 84
– LSDA, LDA, GGA 84
– meta-GGA, dispersion correction 85
excitation error 153
extraframework anion 170–177

ferropericlase 129, 135, 141
fingerprint function 33
five-fold symmetry 110–111
fluorescence 17–18
fluorkyuygenite 182, 184
fluormayenite 182–184

garnet 5, 22, 128–133, 169–170, 181, 185–187
gemology 22
geobarometry 140–141
geometry optimization 80–81
group theory 109

half-metal 79, 97–98
hannayite 66
Hartree-Fock method 83–85
hazenite 66
hexacelsian 56
hierarchical depth 51–52
hierarchical span 51–52
high pressure 55–56, 82, 116, 123, 130–131, 135, 140
high temperature 54–55, 82, 121, 132, 135, 158, 180, 185
highly reducing environment 116
high-resolution narrow scans 10, 16
HTML 15–25
human calcified tissues 159–161
human chromosome 66
hydrogarnet 133, 169–170, 181, 185–187
hydrous ringwoodite 133, 135, 137–143
hydroxyapatite 155, 159–161
– human hydroxyapatite 155, 159–161
– structure determination 160–161
– structural models 160
hydroxylation 92, 176–177, 182–183, 186–187

icosahedral symmetry 110–112, 116
icosahedrite 114–123
icosahedrons 114–115
IMA 128, 129, 184
IMA-List 3–4, 7–9, 14, 24
impact shock 119–120
incommensurability 110
Inductively Coupled Plasma Mass Spectrometry (ICP-MS) 6
information density 38–39
infrared spectroscopy 2–3, 12–13, 133

interaction potential 76
– Morse potential 82
– reactive force field 82
– Coulomb interaction 81
– Lennard-Jones potential 81
inversion center 19
ionic surfaces 78–79

jarosite 50, 59–60

kalsilite 43
Kamchatka 113–116
kaolinite 50, 99, 186
katoite 169–172, 186–187
Khatyrka 113, 116, 119, 121
khatyrkite 113–114, 121
kieserite 50
kimberlite 141
kinetic Monte Carlo 93
kokchetavite 55–56
Kolmogorov complexity 39
kornelite 59–60
kumdykolite 55

lamellae 127–128
laser power 15
laser wavelength 16, 23
Laser-Induced Break-down Spectroscopy (LIBS) 2, 24
lattice periodicity 109
law of rational indices 110
leucite 43, 50, 181
Listventovyi stream 116, 120
Loewenstein's rule 178
lopezite 36–39
lower mantle 128
low-resolution wide scans 10
luminescence peaks 17, 19

magnetite 96–98
majoritic garnet 128–129, 131–132
Mars 2, 23–25, 58
materials engineering 109, 123
mayenite 169–185
– supergroup 182
– type structure 170–177
melanterite 59–60, 66
metallic aluminum 113, 116
metamictization 20
metastability 57
MgO_6 131–132
Mg-wadalite 187–190
Milky Way 122
mica 50, 57
Mindat.org 24
mineral evolution 24, 65–67
mineral samples 2–3, 5
mineral species 1–3, 5, 11, 19, 24, 50, 67
modulated structures 110–111, 159
Monte Carlo 80, 93
Mössbauer 2, 24, 133
muscovite 50, 57
MySQL 14–15

nanomaterials 100, 151–162
NASA 2, 23–25
natrosilite 39
natural quasicrystals 112–123
nebular gas 118
nepheline 43, 50, 118
Newton equation of motion 80
nomenclature 3, 23, 182–184
non-ideal conditions 112

O^- 180–181
O^{2-} 180–181
O-H 10–11
olivine 24, 50, 114, 118, 127–131, 134, 142
Open Data Repository 24–25
open-source 14
orthorhombic 55,113, 129, 157
orthosilicates 129
oxygen isotopes 118
oxygen radicals 180–181

Pandey reconstruction 92
parabutlerite 59–60
parsimony 32–34
paulingite 34, 50–51, 66
Penrose tiling 111–112
pentacalcium-trialuminate 169
periodic boundary conditions 80, 86
periodic table 4, 7, 14, 85
pertlikite 66
phase transformations 23, 54, 59, 65, 128
phason strain 114–115
PHP 14–15
phyllosilicates 65, 98–99, 100, 155

phys-sorption 93
planet formation 123
polymorph 14, 48, 54–55, 58, 75, 116–117,128–132, 157–161
Post Hartree-Fock methods 83
– configurational interaction 83
– Møller–Plesset perturbation theory 83
powder X-ray diffraction 7–9, 113, 128, 149, 158, 171
proteins 162
protophase 185–187
pseudomorph 185–187
pseudopotential 85–86
– norm conserving 86
– ultrasoft 86
pushcharovskite 66
pyrite 58–60, 94–96
pyrophyllite 99–100

quasicrystal 109–124
Queensland 127
quenstedtite 59–60
quotient graph 36–38

radioactive decay 20, 99, 187
Raman active modes 11
Raman spectra 2, 10–11, 15, 20–23
Raman spectral database 2–3
Raman spectroscopy 2, 10, 15–21, 159
rapid cooling 55, 57, 121, 123
rapid quenching 55, 114
reconstruction 77
– matrix notation 77
– Wood's notation 77
records 1, 5, 11, 14, 21, 25
reference library 13
reference spectra 2, 10, 22
relaxation 77, 79, 90–94, 97–99
remnant pressure 140
richetite 66
ringwoodite 127–144
roemerite 59–60
rotational symmetry 109–110, 113
rozenite 59, 60
RRUFF 1–25
– database 2–3, 5
– ID 5, 14
rumpling 77, 92
rutile 33, 92–94

sample collection 5
sample orientation 11
satellite 110
schwertmannite 66
search/match 2, 8, 10–12
seismic discontinuities 128
semiconductor 91–92, 96, 98, 181
semimetal 92, 97–98
serpentine 56, 186–187
Shannon formula 34–35, 38
siderotil 59–60
single-crystal X-ray diffraction 7–8, 113, 115, 149, 171–172
sodalite 46, 179
solar system 1, 109, 118–123
space group 32, 38, 40, 49, 129, 158–161, 169–171
spectroscopy 2, 24, 75, 133, 140, 150, 159, 161
sponge primmorphs 154–156
– sample preparation 154
– inorganic nanocrystals in sponge 155–156
stability 55–57, 61, 75, 78–79, 87, 93–94, 96–97, 109, 123, 128–129, 134, 141, 153
stability field 128–129, 134, 141
state of matter 112
steinhardtite 119–121
stishovite 54, 116–117, 123, 131
structural cages 170–187
structural hierarchy 50–53
structure
– determination 150, 153–154, 157–158, 160, 162
– distortion 77, 98, 131, 170–177
– refinement 49, 153–154, 158, 160, 162, 173, 175
struvite 65
suanite 39
superspace 110–112
surface acidity 90
surface energy 77, 79, 87–88, 92–93, 96–97
svyatoslavite 48, 55, 69, 70
syngas 180–181
synthesis 112, 133, 149, 169, 179, 181, 187–189, 191
szomolnokite 50, 59–60

Tenham meteorite 127
thermal effects 15
thermal expansion 84, 123, 135, 139–141

thermodynamic ensemble 80
thermodynamic integration 89–90, 99
TOC 179–180
TO-layer 99
TOT-layer 99
transition zone 128–129, 132, 141–144
transmission electron microscopy 113, 133, 150
– aberration-corrected TEM 162
– bright-field TEM 150, 160, 162
– environmental TEM 162
– high resolution TEM (HRTEM) 150–151, 155, 157, 159
– scanning- (STEM) 150, 152–155, 157
– (TEM) 150–153, 155–162
tridymite 43, 48, 54
tristetrahedral 185–186

ultrasonic interferometry 135–136
upper mantle 128, 132
URL 14
UV spectroscopy 24

vandendriesscheite 66
vaterite 154, 157–159
– polytypism 159
– structural models 157–159
– structure determination 157
– synthesis 154
Verwey transition 96, 98
vesuvianite 49–50, 185
vivianite 65
voltaite 59–60, 66

wadalite 182, 184, 186–191
wadsleyite 128–134, 141–144
wairakite 61–63
wave function 82–83, 86
website 3, 6, 8, 12, 14, 21
whitlockite 66

XPOW 8–9, 21
X-ray diffraction 1–15, 21, 24, 75, 109, 113, 115, 123, 133, 135-136, 138, 140, 149, 157, 169, 171
– single crystal 7–8, 113, 115, 149, 171–172
– powder (XRPD) 7–9, 113, 128, 149, 158, 171
XtalDraw 8, 21

zeolites 40, 49, 51, 57, 61–63, 66, 71–72, 153, 178–179

Bei Fragen zur Produktsicherheit wenden Sie sich bitte an:
If you have any questions regarding product safety,
please contact:

Walter de Gruyter GmbH
Genthiner Straße 13
10785 Berlin
productsafety@degruyterbrill.com